DATE DUE

MAR 0 5 2011	

COLD

COLD

Adventures in the World's Frozen Places

BILL STREEVER

(L)(B)

Little, Brown and Company
New York   Boston   London

Little, Brown and Company
Hachette Book Group
237 Park Avenue, New York, NY 10017
Visit our Web site at www.HachetteBookGroup.com

First Edition: July 2009

Little, Brown and Company is a division of Hachette Book Group, Inc.
The Little, Brown name and logo are trademarks of Hachette Book Group, Inc.

Library of Congress Cataloging-in-Publication Data
Streever, Bill.
    Cold : adventures in the world's frozen places / Bill Streever.
       p. cm.
    Includes bibliographical references.
    ISBN 978-0-316-04291-8
    1. Arctic regions—Description and travel.   I. Title.
    G608.S69 2009
    910.911—dc22   2008045350

10 9 8 7 6 5 4

RRD-IN

Printed in the United States of America

For my son, Ishmael, who never thinks
it is too cold to play outside

I think myself obliged to give my Readers an account.... Why I thought fit to write of Cold at All?...The subject I have chosen is very noble.

—Robert Boyle, *New Experiments and Observations Touching Cold*, 1683

The captain had been telling how, in one of his Arctic voyages, it was so cold that the mate's shadow froze fast to the deck and had to be ripped loose by main strength. And even then he got only about two-thirds of it back.

—Mark Twain, *Following the Equator*, 1897

CONTENTS

CONTENTS

NOVEMBER

Skis and skiing, a trail closed by a late-season bear, and
freezing trees releasing a burst of heat and flushing the
fluid from their cells

DECEMBER

Overheating in the depths of winter, shadows of Weddell seals
in the sea ice, and Japanese ama divers in water cold enough
to kill most humans

JANUARY

Weather patterns that cause frigid conditions, medieval
weather forecasters burning at the stake, and
a frozen ocean

FEBRUARY

Plummeting temperatures, the cooling of Westminster
Abbey, and approaching absolute zero and the death of matter

MARCH

A search for polar bear dens near forty below zero, winter
apparel, igloos, quinzhees, and a house instrumented to
measure cold

APRIL

Frost-heaved roads, broken pipes, crops destroyed by frost,
and 143 caribou killed by an avalanche

CONTENTS

MAY

JUNE

AUTHOR'S NOTE

Throughout this book, degrees Fahrenheit are used unless noted otherwise because this is the temperature scale most familiar to readers in the United States. Forty below zero Fahrenheit is equal to forty below zero Celsius, and thirty-two degrees Fahrenheit is equal to zero Celsius.

PREFACE

The world warms, awash in greenhouse gases, but forty below remains forty below. Thirty degrees with sleet blowing sideways is still thirty degrees with sleet blowing sideways. Cold is a part of day-to-day life, but we often isolate ourselves from it, hiding in overheated houses and retreating to overheated climates, all without understanding what we so eagerly avoid.

We fail to see cold for what it is: the absence of heat, the slowing of molecular motion, a sensation, a perception, a driving force. Cold freezes the nostrils and assaults the lungs. Its presence shapes landscapes. It sculpts forests and herds animals along migration routes or forces them to dig in for the winter or evolve fur and heat-conserving networks of veins. It changes soils. It preserves food. It carries with it a history of polar exploration, but also a history of farming and fishing and the invention of the bicycle and the creation of Mary Shelley's *Frankenstein*. It preserves the faithful in vats of liquid nitrogen from which they hope one day to be resurrected.

Imagine July water temperatures of thirty-five degrees. Imagine Frederic Tudor of Boston shipping ice from Walden Pond to India on sailing ships in 1833. Imagine Apsley Cherry-Garrard on his search for penguin eggs at seventy below zero in 1911. Imagine a dahurian larch forest that looks like a stand of Christmas trees on Russia's Taymyr Peninsula at sixty below or a ground squirrel hibernating until its blood starts to freeze and then shivering itself back to life.

But none of this is imaginary. Our world warms, but cold remains. In the ordinary passing of a calendar year, the world of cold emerges. For someone with Raynaud's disease, a September stroll temporarily changes cold hands into useless claws. Caterpillars freeze solid in October and crawl away in April. Average temperatures in certain towns drop to twenty below zero in January.

It is time to enjoy an occasional shiver as we worry about a newly emerging climate likely to melt our ice caps, devour our glaciers, and force us into air-conditioned rooms. It is time to embrace and understand the natural and human history of cold. Even in a warming world, a world choked by carbon dioxide and methane, cold persists, biting my lungs and at the same time leaving me invigorated, alive and well on an Arctic spring afternoon with the sun hovering low over an ice-covered horizon and the thermometer at forty below.

COLD

JULY

It is July first and fifty-one degrees above zero. I stand poised on a gravel beach at the western edge of Prudhoe Bay, three hundred miles north of the Arctic Circle, and a mile of silt-laden water separates me from what is left of the ice. The Inupiat—the Eskimos—call it *aunniq,* rotten ice, sea ice broken into unconsolidated chunks of varying heights and widths, like a poorly made frozen jigsaw puzzle. A few days ago, the entire bay stood frozen. During winter, it is locked under six feet of ice. Trucks drive on it to resupply an offshore oil production facility. If one were insane, or if one were simply too cheap to fly, or if boredom instilled a spirit of adventure, one could walk north to the North Pole and then south to Norway or Finland or Russia. Temperatures would range below minus fifty degrees, not counting windchill.

But even in summer, the weather resides well south of balmy. A chill gust runs through me as I stand shirtless on the water's edge, wearing nothing but swimming shorts in the wind and rain.

"The only way to do this," I tell my companion, "is with a single plunge. No hesitation."

I go in headfirst. The water temperature is thirty-five degrees. I come up gasping. I stand on a sandy bottom, immersed to my neck. The water stings, as if I am rolling naked through a field of nettles. I wait for the gasp reflex to subside. My skin tightens around my body. My brain—part of it that I cannot control—has sent word to the capillaries in my extremities. "Clamp down," my brain has commanded, "and conserve heat." I feel as if I am being shrink-wrapped, like a slab of salmon just before it is tossed into the Deepfreeze.

My companion, standing on the beach, tells me that I have been in the water for one minute. My toes are now numb.

Time passes slowly in water of this temperature. I think of the ground, permanently frozen in this region to a depth of eighteen hundred feet. I think about hypothermia, about death and near death from cold. I think of overwintering animals. I think of frozen machinery with oil as thick as tar and steel turned brittle by cold. I think of the magic of absolute zero, when molecular motion stops.

After two minutes, I can talk in a more or less normal tone. But there is little to discuss. There is, just now, almost no common ground between me and my companion, standing on the beach. I feel more akin to the German soldiers whose troop carrier foundered, dumping them into Norwegian coastal waters in 1940. Seventy-nine men did what they could to stay afloat in thirty-five-degree water. All were pulled alive from the water, but the ones who stripped off their clothes to swim perished on the rescue boat. Suffering more from hypothermia than those who had the sense to stay clothed, they succumbed to what has been called "afterdrop" and "rewarming shock." Out of the water, they reportedly felt well and were quite able to discuss their experience. But as the cold blood from their extremities found its way to their hearts, one after another they stopped talking, relaxed in their bunks, and died.

"Three minutes," my companion tells me.

I am a victim of physics. My body temperature is moving toward a state of equilibrium with this water, yielding to the second law of thermodynamics. I shiver.

Several hundred miles southwest of here, six days before Christmas in 1741, the Danish navigator Vitus Bering, employed by Russia, lay down in the sand and died of scurvy and exposure, while his men, also immobilized by scurvy, cold, and fear, became food for arctic foxes. Some accounts hold that Bering spent his last moments listening to the screams and moans of his dying men. The Bering Sea, separating Russia and Alaska, was named for him, and the island where he died, nestled on the international date line, is known today as Bering Island.

Northeast of here, in 1881, Adolphus Greely led twenty-five men to the Arctic, stopping at Ellesmere Island. For most of them, the trip was a slow death that combined starvation, frostbite, and hypothermia. Greely himself, with five others, survived. He eventually took charge of what would become the National Weather Service, where he failed to predict a blizzard in which several hundred people died from frostbite and hypothermia. Many of the dead were schoolchildren.

Half a century after Greely's expedition, in the 1930s, the missionary ascetic Father Henry lived at Pelly Bay, in Canada's Northwest Territories, well above the Arctic Circle. By choice, he resided in an ice cellar. Indoor temperatures were well below zero. The natives would not live in an ice cellar, which was designed to keep game frozen through the short Arctic summer. It was the antithesis of a shelter, analogous to living in a shower stall to avoid the rain. Father Henry believed that it focused his mind on higher matters. Almost certainly, some of the natives believed that Father Henry was mad.

"Four minutes," my companion calls. The stinging in the skin of my thighs has turned to a burning pain. Frostbite is not a real possibility at this temperature, and true hypothermia is at least ten minutes in the future. What I feel is no more than the discomfort of cold.

Frogs are not found this far north, but at their northernmost limit, a few hundred miles from here, they overwinter in a frozen state, amphibian Popsicles in the mud. Frogsicles. But caterpillars are found near here. I sometimes see them crawling across the tundra, feeding on low-growing plants. They freeze solid in winter, then thaw out in spring to resume foraging between clumps of snow. They are especially fond of the diminutive willows that grow in the Arctic.

Ground squirrels overwinter underground. They are related to gray and red squirrels and to chipmunks, but in appearance they are more similar to prairie dogs. In their winter tunnels, their body temperature drops to the freezing point, but they periodically break free from the torpor of hibernation, shivering for the better part of a day to warm themselves. And then they drift off again into the cold grasp of hibernation. Through winter, they cycle back and forth—chill and shiver, chill and shiver, chill and shiver—surviving.

Arctic soil behaves strangely around the hibernating ground squirrels. Underground, liquid water is sucked toward frozen water, forming lenses of almost pure ice. The soil expands and contracts with changing temperatures, forming geometric shapes, spitting out stones on the surface, cracking building foundations. Wooden piles cut off at ground level are heaved upward by ground ice, sadly mimicking a forest in this frozen treeless plain.

This water I stand in feels frigid, bitingly cold, but in the greater scheme of things it is not so cold. A block of dry ice—frozen carbon dioxide—has a surface temperature just warmer than minus 110 degrees. James Bedford has been stored in liquid nitrogen at minus 346 degrees since 1967, awaiting a cure for cancer. The surface of Pluto stands brisk at minus 369 degrees. Absolute zero is some five hundred degrees colder than the water that surrounds me.

"Five minutes," my companion tells me. I leave the water, shivering, my muscles tense. It will be two hours before I feel warm again.

❄   ❄   ❄

There is more than one way to measure temperature. Daniel Fahrenheit, a German working in Amsterdam as a glassblower in the early 1700s, developed the mercury thermometer and the temperature scale still familiar to Americans. He built on work dating back to just after the time of Christ and modified by the likes of Galileo, who used wine instead of mercury, and Robert Hooke, appointed curator of the Royal Society in 1661, who developed a standard scale that was used for almost a century. In 1724, Fahrenheit described the calibration of his thermometer, with zero set at the coldest temperature he could achieve in his shop with a mixture of ice, salt, and water, and 96 set by sticking the instrument in his mouth to, in his words, "acquire the heat of a healthy man." He found that water boiled at 212 degrees. With only a minor adjustment to his scale, he declared that water froze at 32 degrees, leaving 180 degrees in between, a half circle, reasonable at a time when nature was believed by some to possess aesthetic symmetry.

Anders Celsius, working in Sweden, came up with the Celsius scale in 1742. Conveniently, it put freezing water at zero and boiling water at one hundred degrees. Less conveniently, it set in place a competition between two scales. An Australian talking to an American has to convert from Celsius to Fahrenheit, or the American will think of Australia as too cold for kangaroos. An American talking to an Australian has to convert from Fahrenheit to Celsius, or the Australian will think of America as too hot for anything but drinking beer. The Australian is forced to multiply by two and add thirty-two, or the American is forced to subtract thirty-two and divide by two. Or, as more often happens, they drop the matter of temperature altogether.

Lord Kelvin realized in 1848 that both Fahrenheit and Celsius had set their zero points way too high. He understood that heat could be entirely absent. At least conceptually, absolute zero was a possibility. He came up with his own scale, based on degrees Celsius, but with zero set at the lowest possible temperature, the point at which there is no heat. Zero Kelvin is 459 degrees below zero Fahrenheit. Just above this temperature, helium becomes a liquid. Anywhere close to absolute zero, and all things familiar to the normal world disappear. Molecular motion slows and then stops. A new state of matter, called a "super atom"—something that is neither gas nor liquid nor solid—comes into being. But Kelvin's understanding of the strange world that exists within a few degrees of absolute zero was theoretical. By the time he died, in 1907, his colleagues were struggling to force temperatures colder than 418 degrees below zero, 41 degrees above absolute zero, and helium had not yet been liquefied.

One of the physicists who first achieved a temperature low enough for the formation of a super atom, which did not occur until 1995, had this to say: "This state could never have existed naturally anywhere in the universe, unless it is in a lab in some other solar system."

Our planet's polar explorers used, for the most part, Fahrenheit's scale, but rather than talking of degrees below zero, they often talked of "degrees of frost." One degree of frost was one degree below freezing Fahrenheit. An explorer might write in his journal of fifty degrees of frost—eighteen degrees below zero Fahrenheit—and in the next paragraph tell of the amputation of a frozen toe, or describe himself gnawing on a boot to stave off the starvation that so often accompanies cold, or mention in passing how he had to beat fifteen pounds of ice from the bottom of his sleeping bag before bedding down for the night. Or, after an especially cold and uncomfortable spell, he might write of the relative warmth and relief of fifty-five degrees below zero Fahrenheit. Apsley Cherry-Garrard, who supported Robert Falcon Scott on his disastrous 1910 Antarctic expedition, did just that. "Now," he wrote, "if we tell people that to get only 87 degrees of frost can be an enormous relief they simply won't believe us." But an enormous relief it would be for one accustomed to camping at 75 degrees below zero, or 107 degrees of frost.

In his memoirs, Cherry-Garrard concurred with Dante, who placed the circles of ice beneath the circles of fire in his vision of Hell.

❄   ❄   ❄

It is July eighteenth and nearly fifty degrees under an overcast sky. I walk slowly across Arctic tundra next to an abandoned airstrip, stalking *Gynaephora rossii*. The trouble with this beast—the woolly bear caterpillar of the far north—is that it is not easy to find here near Prudhoe Bay. Woolly bear caterpillars are substantially smaller than woolly mammoths. Woolly mammoths, but for their unfortunate extinction, would be easy to spot. But these woolly bear caterpillars are smaller than a mammoth's eyebrow. And this terrain is not conducive to stalking insect larvae. Though flat and treeless, the terrain is uneven. The prudent searcher watches his footing when he should be watching for caterpillars. Every few steps, water-filled cracks in the ground require minor detours. The cracks form when the ground contracts and expands in response to temperature changes. Once a crack forms, it fills with water. When the water freezes, the ice expands and widens the crack. A wedge of ice forms and grows, and the crack eventually becomes too wide to step across. Cracks intercept other cracks. Together, they make a network outlining polygons that are thirty feet wide. They polygonize the landscape.

The cracks are beginning to find their way across the abandoned airstrip. Next to the cracks, where water pools through the summer months, grass grows lusciously green. Between the cracks, in the centers of the polygons, the greenery struggles—less dense, less luscious. Or even not luscious at all. Despite all this water, the ground can be dry between the cracks, and dust covers some of the plants. Just days ago, the creamy flowers of arctic dryas made patches of this dry ground look like miniature gardens of snowy roses. Now their dried scraggly puffball seed heads are all that remains. In the Arctic, blink, and summer is gone.

Underneath, eighteen inches down, the ground is frozen. It remains frozen for a third of a mile before heat from the earth's innards overcomes the cold from above. Poking the ground with a steel rod, one can feel the permafrost—the permanently frozen ground. It's like hitting bedrock just eighteen inches down.

Where are the caterpillars? I find a biologist who has been working here since May, counting birds. I ask her if she has seen any caterpillars. "I've only seen one," she tells me.

Later, I talk to an Inupiat elder. "I see them sometimes," he says. "Maybe once each year." Inupiat frequently pause when they talk, leaving what might seem like an uncomfortable silence. I have been told that the pauses give them time to think and therefore to avoid the mindless patter of whites. "They like high ground," he says after a moment. "I see them near my camp at Teshekpuk Lake."

The little beasts eat willow buds. I squat on the tundra to check some of the willows growing on the high ground between water-filled cracks. These willows are related to the taller willows of warmer climates, but they never stand more than a few inches tall. Their trunks can be measured in fractions of an inch. I find neither caterpillars nor gnawed buds. I pluck a leaf and pop it into my mouth. It tastes like an aspirin salad. I move on.

Hyperactive birds fly around the airstrip. A plover screeches at me and makes threatening dives, driving me away from its young. In tundra ponds and in water-filled cracks, phalaropes swim in tight circles, their heads bobbing as if connected to their feet. A pair of snow buntings perch for a second on top of a pipeline next to the

airstrip and then fly off. A long-billed dowitcher, its beak dispropor-
tionately long, flushes from the ground in front of me. Behind it, a
hundred yards away, five caribou graze, their antlers imitating the
beak of the dowager in their freakish length.

Soon all of this activity will cease. The birds will fly away. The
caribou will march south. The caterpillars will simply freeze. That
is why I am interested. That is why I want one of these caterpillars. The little devils have figured out how to freeze solid without
dying. They are slow growers. It might take a decade before they are
ready to metamorphose into grayish moths. That means they survive through ten winters here in the Arctic. When spring comes, they
thaw and go back to eating. For a pet lover who travels, they could be
the perfect solution. Cute, furry, and quiet, and the freezer serves as
a kennel. But where are they? If I were looking for oil, I would have
just successfully drilled a dry hole, a duster. I have been skunked by
a caterpillar.

❄   ❄   ❄

The polar explorers were great keepers of journals, and many of the
survivors produced memoirs. Cold for the polar explorers came with
a sense of pride, but also uncertainty, hunger, exhaustion, and death.
The body's boilers run on food, and as often as not, death from pro-
longed exposure to cold combines starvation, frostbite, and hypothermia. When one reads past the stoicism and heroics, the history
of polar exploration becomes one long accident report mixed with
one long obituary.

There was, of course, discomfort. In 1909, Ernest Shackleton
traveled to within ninety-seven miles of the South Pole. Realiz-
ing that his provisions would be stretched if he pushed farther, he
turned around. He told his wife, "I thought you would rather have
a live donkey than a dead lion." In 1914, during a later exploration,
his ship *Endurance* was iced in and eventually abandoned. He led

his men slowly across the ice. In his travelogue, he wrote, "I have stopped issuing sugar now, and our meals consist of seal-meat and blubber only, with 7 ozs. of dried milk per day for the party." This is at a time of inactivity, camped on ice. "The diet suits us, since we cannot get much exercise on the floe and the blubber supplies heat," he wrote. Eventually, the ice gave way, cracking under his camp. "The crack had cut through the site of my tent," he wrote. "I stood on the edge of the new fracture, and, looking across the widening channel of water, could see the spot where for many months my head and shoulders had rested when I was in my sleeping bag."

Charles Wright survived Robert Falcon Scott's 1910 Antarctic expedition and knew just how important those sleeping bags were. He—with Apsley Cherry-Garrard, who had grasped Dante's reasons for placing the circles of ice beneath those of fire in the depths of Hell—was one of the men who supported Scott, hauling Scott's gear south for the first leg into the heart of Antarctica. The support team turned back and waited at their base camp, but Scott and the four men who continued to the pole would not survive. Long afterward, at eighty-six years old, Wright talked to an interviewer about man-hauling sleds in Antarctica. The interviewer asked about toilet habits on the trail, the point being that getting up in the middle of the night to relieve oneself involved more than just stepping outside of the tent in your boxers. "You see," Wright explained,

you've come from your sleeping bag, you've taken into the sleeping bag all the frozen sweat of the previous day, and the previous day and the previous day and the previous day. There's a log of it at the end. And during the night you first melt that frozen sweat. And very often it freezes at the bottom of the bag, where your feet are. And if you're going to have a decent night you've got to melt all that before you have a chance. And even then it's not comfortable because whatever is next door is wet and cold, and every breath you take brings some of the cold stuff into the small of your back. So a winter's night when you're sledging is not a comfortable thing at all. But you've got to, before you get anywhere, you've got to melt the ice. And sometimes there's fifteen pounds of ice or something like that that's got to be turned into water before you begin to sleep.

From Wright's account it is clear that Antarctic explorers disciplined their bladders and stayed in those half-frozen bags as long as possible.

Scott himself kept a journal right up until his death. Eight months later, a search party found his camp. In the camp, Scott's frozen body lay between two of his frozen companions. The three men in the tent, it has been said, looked as if they were sleeping. The three bodies, along with Scott's journal, were recovered.

Scott's journal records noble behavior and tragedy. By the middle of January 1912, eager to be the first to reach the South Pole, Scott and the four men who went with him stumbled on sled tracks and camps left by the Norwegian explorer Roald Amundsen, who had beaten them to the pole by four weeks. Scott's party pressed on to the pole anyway. "Great God!" Scott wrote, "this is an awful place and terrible enough for us to have laboured to it without the reward of priority." Disappointed, the men struggled back toward their base camp.

"Things steadily downhill," Scott wrote in early March. "Oates' foot worse. He has rare pluck and must know that he can never get through. He asked Wilson if he had a chance this morning, and of course Bill had to say he didn't know. In point of fact he has none." Later, Oates, recognizing that he was slowing the party and endangering their lives, talked to his companions. "I am just going outside and may be some time," he said. Afterward Scott wrote, "He went out into the blizzard and we have not seen him since."

Scott wrote about himself, "My right foot has gone, nearly all the toes — two days ago I was proud possessor of best feet. These are the steps of my downfall. Like an ass I mixed a small spoonful of curry powder with my melted pemmican—it gave me violent indigestion. I lay awake and in pain all night; woke and felt done on the march; foot went and I didn't know it. A very small measure of neglect and have a foot which is not pleasant to contemplate. Amputation is the least I can hope for now, but will the trouble spread?"

Later he wrote, "It seems a pity, but I do not think I can write more." And after this, he had but one final entry: "For God's sake look after our people." Scott ended his days eleven impossible miles from a supply depot that would have saved his life.

Frostbite is a common theme among polar explorers. Captain George E. Tyson was marooned with his crew on an Arctic ice floe in the winter of 1872 and spring of 1873. "The other morning," he wrote, "Mr. Meyers found that his toes were frozen—no doubt from his exposure on the ice without shelter the day he was separated from us. He is not very strong at the best, and his fall in the water has not improved his condition."

Food, or a lack of food, is another common theme. Roald Amundsen, when he beat Robert Falcon Scott to the South Pole, used sleds pulled by dogs. The dogs doubled as a food supply. Amundsen had this to say about men who pursued their destinies at the poles: "Often his search is a race with time against starvation."

Robert Flaherty published the story of Comock, an Inuit. In the narrative, Comock explains how he and his family lived on Mansel Island in the Canadian Arctic early in the twentieth century. They were on the island alone, isolated for ten years from their extended families and the villages that dotted the Arctic. They were at times well fed.

"Look at our children," Comock's wife said to Comock. "They are warm."

And Comock, in his narrative, added, "There were little smokes rising from the deerskin robes under which they slept."

But later, food became scarce. "We shared with our dogs the dog meat upon which we lived," Comock reported. One of his companions said that seal meat offered warmth, while dog meat did not. Comock feared the dogs would eat the children.

Frederick Cook, who probably reached or at least came close to the North Pole in April 1908, almost a full year before Robert E. Peary, ran into trouble and could not return to civilization quickly enough to defend himself against Peary's own claim and what has been described as Peary's slander. Like Apsley Cherry-Garrard, Cook concurred with Dante, but with more drama and self-aggrandizement: "We all were lifted to the paradise of winners as we stepped over the snows of a destiny for which we had risked life and willingly suffered the tortures of an ice hell." But after two days at the pole, he described a feeling of anticlimax. The pole itself, after all, was just another frozen camp in a frozen landscape. "The intoxication of success was gone," he wrote in his memoir. "Hungry, mentally and physically exhausted, a sense of the utter uselessness of this thing, of the empty reward of my endurance, followed my exhilaration."

And who has heard of Lieutenant George De Long? In an 1879 attempt to reach the North Pole, De Long and twenty men abandoned their ship to the ice. They dragged three small boats across the ice for nearly three months before finding open water. One boat was lost, but two made it to Siberia's Lena River delta. This was early October. Though suffering from frostbite and exhaustion, the men were not complainers. De Long wrote, "The doctor resumed the cutting away of poor Ericksen's toes this morning. No doubt it will have to continue until half his feet are gone, unless death ensues, or we get to some settlement. Only one toe left now. Temperature 18°."

Like Scott, though perhaps with less panache, De Long maintained his journal until the end:

October 17th, Monday.—One hundred and twenty-seventh day. Alexey dying. Doctor baptized him. Read prayers for the sick. Mr. Collins' birthday—forty years old. About sunset Alexey died. Exhaustion from starvation.

October 21st, Friday.—One hundred and thirty-first day. Kaack was found dead about midnight between the doctor and myself. Lee died about noon. Read prayers for the sick when we found he was going.

October 24th, Monday.—One hundred and thirty-fourth day. A hard night.

The next two days contain only the date and the number of days. Then:

October 27th, Thursday.—One hundred and thirty-seventh day. Iversen broken down.

October 28th, Friday.—One hundred and thirty-eighth day. Iversen died during early evening.

October 29th, Saturday.—One hundred and thirty-ninth day. Dressler died during night.

October 30th, Sunday.—One hundred and fortieth day. Boyd and Gortz died during night. Mr. Collins dying.

The bodies of De Long and nine others were recovered the following spring.

❄   ❄   ❄

It is July twenty-sixth and sunny. The mercury rises to fifty-two degrees here on Narwhal Island, ten miles north of Alaska's North Slope. Nothing but water and ice separates me from the North Pole. I have, for the past hour, been taking my jacket off and putting it back on. Each time I take it off, a breeze comes in from the north, from the pack ice, like the draft from an open freezer door. Each time I don my jacket, the door closes and the breeze stops. I watch the ice flow past, a regatta of white and blue abstract sculptures. One could not quite step from one chunk of ice to the next without swimming, but boating just now would be a challenge. Here is an ice chunk the size of a suitcase, there one the size of a small house, several in a row the size of compact cars. The breeze comes from the north, but the ice moves to the west, propelled by currents, the bulk of each chunk hanging underwater like an aquatic sail.

Occasionally, a chunk of ice strands next to the shore, hard aground. Another chunk butts up against the first. They grind. Water drips from their tops continuously. Pieces of ice break off, dropping into the Beaufort Sea with splashes that sound remarkably similar to those produced by bass jumping in a still pond. I wade into the sea, break off a piece of ice, and pop it into my mouth. It tastes as fresh as springwater. The molecules in ice are packed in an orderly fashion, forming crystals. There is little space between the molecules for salt ions.

Farther out, between here and the horizon, the ice is more densely packed and in places continuous. Fog banks hover over the ice like plumes of smoke. Occasionally, maybe once each half hour, the pack ice cracks under the pressure of movement, of collisions, of one body striking another. The cracking sounds like distant cannon fire.

The beach I stand on is a mix of gravel and sand. It looks as though someone has worked it over with a bulldozer. The ice, in places, has plowed the sand into piles, left deep gouge marks on the shore, or dumped moraines of gravel above the tide line. Although I am on the island's northern shore, I can turn and see across the island to the other side, and beyond that the mainland, peppered with oil field facilities. A collapsed wooden shack stands on the island, the remains of a long-forgotten scientific party. A large red buoy stranded near the middle of the island speaks of storm tides or ice overrunning the land. Wandering the island, I see only two species of plants clinging to life in scattered patches. Driftwood has accumulated in bigger patches. The rest seems to be bare sand and gravel.

An arctic tern screams at me, then swoops in, obviously protecting a nest. After four or five swoops, it connects with my hat. It leaves me no choice but to find the nest. I look for a small depression in the sand. The closer I get, the more agitated the bird becomes. It dives closer and closer, screaming "Warmer, warmer, warmer." It backs off a bit, telling me "Colder, colder," and I change direction. I move slowly, taking a careful step, scanning the ground in front of me, then taking another careful step. It swoops at me from behind, but I can see its shadow coming. As I get warmer, the tern gets more aggressive. I duck as its shadow closes in. And there is the nest. This late in the year, the tern has but a single egg to protect. Two feet away, a long-dead chick, its body stiff and its eyes glazed, lies on the sand. I back off, ashamed to have disturbed the nesting tern and its lone surviving egg.

Common eiders nest here, too, in bowls scraped from the sand along the edges of driftwood piles. Their nest bowls are much bigger than those of terns, and they are lined with down — eiderdown, as it turns out, plucked from the breasts of females and prized as the best of down, soft and warm and far better than that of domestic ducks. Most of the eider nests are empty, but a few still hold as many as five pale green eggs, somewhat larger than chicken eggs.

For both the eiders and the terns, these may be second or third nest attempts. It seems late in the year to start a family. By September, the eiders will head for open water, where they overwinter, swimming and feeding. The terns, winter averse, will fly twelve thousand miles to Antarctica and then return next spring. During its life, a tern will travel a distance equivalent to that of a round-trip to the moon.

I migrate back to the island's northern shore, scanning the ice through my binoculars. I hope to see a polar bear or at least a seal, but all I see is ice and water. Despite the island's name, no narwhals frolic here today. Narwhals, with their long tusks, live beneath the ice, coming up to breathe in open leads and holes, and moving toward coastal areas in summer. They are common in the Atlantic sector of the Arctic, well to the east. The rare straggler finds its way to Alaska, but I have not seen one in these waters in the five years I have been coming here. The island, someone tells me later, was named for a nineteenth-century whaling ship rather than the narwhal itself.

The seals, the polar bears, and the narwhals mock me, like woolly bear caterpillars, here yet not here, here yet nowhere to be seen.

❄   ❄   ❄

Adolphus Greely, then a lieutenant in the U.S. Army, led his twenty-five men north in the summer of 1881. They made it past eighty-three degrees, some four hundred miles south of the pole, then turned around. The relief ship intended to pick them up could not pass through the ice. A second relief ship sank. The men froze and starved in the far north.

Before it was over, Greely's men experienced an intense aloneness and ate caterpillars. They also ate leather shoelaces. Sealskin lashings became stew. Sleeping bag covers in the nineteenth century were oiled to render them waterproof, and well boiled, the covers were rendered into broth. The men found crumbs of bread as the snow melted around their camp. They lamented the absence of plants and lichens that in other times would not have been considered fit for consumption. They ate hundreds of pounds of amphipods—sea fleas—using, among other things, the remains of dead comrades as bait. One man ate bird droppings, apparently convinced that undigested seeds that passed through birds' guts would provide sustenance. The men divided the soles of an old pair of boots. Later, there would be accusations of cannibalism.

"Everybody was wretched," Greely wrote later, "not only from the lack of food, but from the cold, to which we are very sensitive." Like other Arctic explorers, his narrative is one of death: "Lieutenant

Kislingbury, who was exceedingly weak in the morning at breakfast, became unconscious at 9 a.m. and died at 3 p.m. The last thing he did was to sing the Doxology and ask for water." The men shared sleeping bags to conserve warmth: "Ralston died about 1 a.m. Israel left the bag before his death, but I remained until driven out about 5 a.m., chilled through by contact with the dead." Greely had one man shot for stealing food.

Sergeant David Brainard kept a journal. Like Greely, he focused on the lack of food and described the deaths of comrades. The cold did not go unnoticed. He wrote of sunbathing at forty degrees in May. He wrote of June temperatures well below freezing. After a storm and a very hard night outside, he wrote of suicide: "Of all the days of suffering, none can compare with this. If I knew I had another month of this existence, I would stop the engine this moment."

Others had similar thoughts. "Schneider," Brainard wrote, "was begging hard this evening for opium pills that he might die easily and quickly."

It may go without saying that some of the men suffered from frostbite.

The absence of food was in a very real sense the result of the climate where they resided. The Arctic, despite seasonal and regional abundances of seals and whales and even caribou, is often desolate. And the cold forces one to eat more, to burn more fuel, further compounding the scarcity of food. In a scientific paper written in 2002, it was estimated that the Greely expedition was two million calories short of minimum survival rations. A diet of six thousand calories per day is not unusual for explorers in the polar regions. The calories are needed for both warmth and activity. This is as true for animals as for humans. Bird feathers, when oiled, can no longer keep a bird's skin dry. For a while, an oiled bird will shiver to maintain its body temperature, but shivering requires food. Hypothermic birds die of starvation compounded by hypothermia, or hypothermia compounded by starvation. The same thing killed most of Greely's men.

Winfield Schley, who commanded the boat that picked up seven Greely expedition survivors in 1884, saw Greely through an opening in what was left of a tent. "It was a sight of horror," Schley wrote. "On one side, close to the opening, with his head toward the outside, lay what was apparently a dead man. His jaw had dropped, his eyes were open, but fixed and glassy, his limbs were motionless. On the opposite side was a poor fellow, alive to be sure, but without hands or feet, and with a spoon tied to the stump of his right arm."

❆　❆　❆

It is July thirty-first. Here, a mile from the Beaufort Sea, the thermometer struggles to break forty degrees. My companion, a botanist specializing in Arctic plants, wears rubber boots, wool socks, trousers made of synthetic wicking material, blue Gore-Tex overpants, a sweatshirt and a light jacket under a green plastic raincoat, gloves, and a fleece-lined hat. The wind blows at something like twenty miles per hour. A blanket of fog, thick and damp, covers everything.

Paul Siple is credited with conceptualizing windchill factors in a report written in 1940 but held as a military secret until 1945. He hung water-filled plastic cylinders from a long pole at the newly established Bay of Whales Antarctica base and developed what became known as the Siple-Passel equation for calculating windchill. The windchill factor quantifies the amount of heat lost to wind combined with cold. It expresses what the temperature feels like when the wind blows. Heat lost to wind increases as the square of the wind's velocity. A day with forty-degree temperatures and twenty-mile-per-hour winds feels the same as a still day at thirty degrees. It gets worse as it gets colder. At twenty-five degrees below zero with a thirty-mile-per-hour wind, it feels like sixty below. A common footnote on windchill charts warns that frostbite will occur within five minutes under these conditions.

Fog makes things worse still. The moisture in the air sucks heat away faster than dry air ever could. Fog chills to the bone. Meteorologists sometimes calculate the apparent temperature by combining the measured temperature, the wind speed, and the humidity. This is sometimes called "relative outdoor temperature." Most days, knowing this is no comfort whatsoever.

I wear rubber boots, cotton socks, jeans, and a light jacket. No overpants. No gloves. No fleece-lined hat. I suffer in the midsummer cold. Clothes make the man, or, at least, clothes make the man warm.

❄   ❄   ❄

Adolphus Greely lived to see his ninetieth birthday. He became the first American soldier to enlist as a private and retire as a general. He commanded the erection of thousands of miles of telegraph wires, many of them in Alaska. He oversaw the relief effort following the 1906 San Francisco earthquake, and he was a founding member of the National Geographic Society. For a time, he ran the Weather Bureau, then part of the Army Signal Corps. He was in charge when the Blizzard of January 1888 swept through middle America.

Greely's bureau issued this prediction: "A cold wave is indicated for Dakota and Nebraska tonight and tomorrow; the snow will drift heavily today and tomorrow in Dakota, Nebraska, Minnesota and Wisconsin."

In places, temperatures dropped eighteen degrees in less than five minutes. In Helena, Montana, the temperature dropped from just over forty degrees to nine below in less than five hours. In Keokuk, Iowa, it dropped fifty-five degrees in eight hours. These temperatures do not include the windchill. They are straight temperatures, read from thermometers. Windchill temperatures were colder than forty below.

When the blizzard was over, people found cattle frozen in place, standing as if grazing, their once hot breath now formed into balls of ice around their heads. A government official estimated that something like 20,000 people were "overtaken and bewildered by the storm." Of these, about 250 died from hypothermia and complications of frostbite. The temperature dropped too far too fast. The snow, blowing sideways, reduced visibility to what is called "zero-zero"—one can see zero feet upward and zero feet sideways. People staggered around blindly outside. Cattle, horses, and people, unable to see but knowing they had to seek shelter, wandered downwind. No amount of food would have helped the victims. They died from the cold alone. Because so many of the storm's victims were children, the blizzard became known as the School Children's Blizzard.

Sergeant Samuel Glenn, based in Huron, South Dakota, working for Greely's Weather Bureau, described the suddenness and severity of the storm:

The air, for about one minute, was perfectly calm, and voices and noises on the street below appeared as though emanating from great depths. A peculiar hush prevailed over everything. In the next minute the sky was completely overcast by a heavy black cloud, which had in a few minutes previously hung suspended along the western and northwestern horizon, and the wind veered to the west (by the southwest quadrant) with such violence as to render the observer's position very unsafe. The air was immediately filled with snow as fine as sifted flour. The wind veered to the northeast, then backed to the northwest, in a gale which in three minutes attained a velocity of forty miles per hour. In five minutes after the wind changed the outlines of objects fifteen feet away were not discernible.

After the blizzard, a farmer named Daniel Murphy went out to his haystack. From inside the haystack, he heard a voice. "Is that you, Mr. Murphy?" The voice belonged to nineteen-year-old Etta Shattuck. She had staggered through the windblown snow and, as a last and only resort, had crawled into the haystack. She stayed there just over three days without food or water. Frostbite came on, as it always does, painlessly. There is a sense of cold and stiffness and numbness, but no pain. By the time flesh reaches a temperature of forty-five degrees, nerve synapses no longer fire. All feeling is gone. And then the tissue freezes. Ice crystals form first between the cells. Because ice excludes salts, the remaining liquid between the cells becomes increasingly salty. Osmosis draws water from within the cells toward the saltier fluid outside the cell walls. The cells become dehydrated. Proteins begin to break down. Ice crystals eventually form inside the cells themselves. The sharp edges of the ice crystals tear cell membranes. The flesh dies, starting with the skin. Usually the first skin to die is that of the fingers or toes or ears or nose. Death moves into the muscles, the veins, the bones. Whole limbs, once lively, freeze solid and are dead.

Etta seems to have crawled into the haystack headfirst. She prayed. She sang hymns. She listened to the wind blow. She shared the haystack with mice. At one point, Etta felt the mice rustling through the stack and even nibbling at her wrists. She later explained that this was comforting rather than terrifying. It told her that she was not alone in the world. Because she had crawled in headfirst, her feet and legs were more exposed than her torso. They froze.

Saved from the haystack, Etta went through two rounds of amputations. The newspapers got wind of her, and for a short time she became something of a hero. The *Omaha Bee* set up "The Shattuck Special Fund."

"Miss Etta Shattuck," a reporter wrote, "the young school teacher who lost both limbs from the exposure in the recent storm, will be incapacitated for any service by which she may derive a living. It is desired that $6,000 be raised." But infection set in. She was nineteen years old when she was caught in the blizzard, and she died without seeing her twentieth birthday.

Never mistake frostbite for hypothermia. Frostbite freezes extremities, while hypothermia cools the body's interior. Humans function best at a core temperature of just under ninety-nine degrees. At windchills of minus forty degrees, with serviceable clothing, it is reasonable to expect the core temperature to drop at something like one degree every thirty minutes. When the core drops to ninety-five, significant symptoms appear. People shiver uncontrollably. They become argumentative. They feel detached from their surroundings. As their minds slow, they become what winter travelers sometimes refer to as "cold stupid." They become sleepy.

A thirteen-year-old boy who survived the School Children's Blizzard later recounted his experience. "I felt sleepy," he said. "I thought if I could only lie down just for a few minutes I would be all right. But I had heard the farmers telling stories about lying down and never getting up again in snow storms. So I kept on, but I finally got to the point where I could hardly lift my feet any more. I knew that I couldn't stand it but a minute or two longer."

At a core temperature of about ninety-three degrees, amnesia complicates things. Do we turn right or left? Did I put that glove in my pocket? Have I been here before?

At ninety-one degrees, apathy settles in. Muscles by now are stiff and nonresponsive. If one continues moving at all, one begins to stagger.

When the core temperature reaches ninety degrees, the body's ability to fight the cold diminishes, and the core temperature tumbles downward. The heart itself becomes sluggish. Blood thickens. Lactic and pyruvic acids build up in tissues, further slowing the heartbeat.

It is possible to survive core temperatures as low as eighty-seven degrees, but only with rescue and rewarming. At this temperature, self-rescue is almost impossible. Hallucinations are common. The mind imagines warm food and dry sleeping bags. The ears might hear music. A survivor might report looking down from above on his own struggling body, or he might remember strolling away from his own prone carcass in the snow. Victims at this point have crossed the line between cold stupid and what is sometimes called "cold crazy."

Just shy of death, victims may experience a burning sensation in the skin. This may be a delusion, or it may be caused by a sudden surge of blood from the core reaching the colder extremities. The last act of many victims is the removal of their clothes—the ripping away of collars, the disposal of hats. Doctors sometimes call this "paradoxical undressing."

A Nebraska newspaper explained why some victims of the School Children's Blizzard were missing clothes. "At this stage of freezing strange symptoms often appear: as the blood retires from the surface it congests in the heart and brain; then delirium comes on and with it a delusive sensation of smothering heat. The victim's last exertions are to throw off his clothes and remove all wrappings from his throat; often the corpse is found with neck completely bare and in an attitude indicating that his last struggles were for fresh air!"

During the School Children's Blizzard, a seventeen-year-old girl froze to death standing up, leaning against a tree.

Nebraska teacher Lois Royce wandered through the blizzard with two nine-year-old boys and a six-year-old girl. They could not find shelter. The girl was calling for her mother, begging to be covered up. The boys died. The girl lasted until daybreak. Lois eventually crawled to the safety of a farmhouse.

Johann Kaufmann, a farmer, found his frozen children after the storm. "Oh God," he cried out, "is it my fault or yours that I find my three boys frozen here like the beasts of the field?" The bodies were frozen together. They had to be carried back to the cabin as one and thawed before they could be separated.

In Laura Ingalls Wilder's *Little House on the Prairie,* the horror of blizzards is discussed by children. Laura is asked what she would do if caught in a blizzard. "I wouldn't get caught," she answers.

And Emily Dickinson, in "After Great Pain," seems to have thought of hypothermia:

As Freezing persons, recollect the Snow —
First — Chill — then Stupor — then the letting go —

AUGUST

It is August second and sixty degrees. I watch a fisheries biologist wade into the Beaufort Sea. On and off, he has been wading into the Beaufort Sea for more than twenty years, collecting fish as part of a long-term study. He wears chest waders, but the cold soaks right through. Even when he stays dry, the plastic fabric presses against his skin, feeling wet. At their best, waders in cold water give meaning to the word "clammy." And at times the waders leak, or they are overtopped by a wave, or he steps into a hole and they fill up with ice water.

I sit in the Zodiac as he boards from water that reaches close to the top of the waders. He rolls across the edge of the Zodiac, leaning into the boat and straightening his legs, so that his feet are higher than his head. Water drains from the waders into the boat.

I tell him about the book I am writing. I tell him of my five-minute bath in the Beaufort Sea. He has this to say: "You should do a book called *Warmth*. You could do all the background research in Aruba."

"What would be the fun in a book on warmth?" I ask. And then it occurs to me: fire walking.

❄   ❄   ❄

Some polar explorers stayed warm. Part of their secret was clothing. Richard Byrd, famous for a failed attempt to fly over the North Pole and for a successful flight over the South Pole, spent many months in Antarctica. It was during a Byrd expedition to Antarctica that Paul Siple hung his water-filled cylinders in the breeze and worked out the principles of windchill. In 1933, strapped for funds, Byrd overwintered alone in Antarctica. "Cold was nothing new to me," he wrote, "and experience had taught me that the secret of protection is not so much the quantity or weight of the clothes as it is the size and quality and, above all, the way they are worn and cared for." At sixty-five below zero, he wore, among other things, a mask. "A simple thing," he wrote, "it consisted of a wire framework overlaid with windproof cloth. Two funnels led to the nose and mouth, and oval slits allowed me to see. I'd breathe in through the nose funnel, and out through the mouth funnel; and when the latter clogged with ice from the breath's freezing, as it would in short order, I brushed it out with a mitten." He wrote of walking comfortably outside, suited up, and he compared himself to a diver. This was in 1933, during a time when divers wore heavy canvas suits with brass and copper helmets bolted to the suits and weighted, metal-framed boots on their feet.

Another secret to warmth involved seeking help from the locals. This worked only in the Arctic, where there were local people from whom to seek help. Isaac Hayes walked away from his ship when it froze into the Arctic ice in 1854. In general, he was scornful of the natives he encountered on his way south, whom he called "Esqui-maux." He thought of them as savages, but he was not above accept-ing their hospitality. After taking refuge in a village, he wrote, "The hut was warmer by 120° than the atmosphere to which we had been so long exposed."

Patience to wait out the cold played a role in survival, too. Fridt-jof Nansen's writings, though they were not intended to do so, make a mockery of the suffering of the likes of Scott and Greely and Bering. Nansen, in 1888, was the first to cross the Greenland ice sheet. He did it on skis. Later, he thought it reasonable to intentionally freeze his boat into the pack ice and let the drifting polar ice carry him across the Arctic. In 1893, he sailed from Norway in the *Fram*, a vessel not much bigger than a large yacht. He traveled with thirteen Norwegians, because, he joked, only Norwegians could tolerate one another for month after month on a boat drifting with the pack ice. A year and a half after freezing in, Nansen and one of his men left the *Fram*. Apparently at least in part out of boredom, they headed north with three sleds, two kayaks, and twenty-eight dogs. After three weeks, Nansen was within four degrees of the pole, a new record, but there he turned. Heading south, the two men overwintered on an island. They dug a hole three feet deep, which would have meant chiseling through permafrost with the consistency of hardened concrete. They put stones three feet high around the hole and then roofed it with walrus hides and snow. They laid in game, mostly bear.

Nansen and his companion gained weight that winter. Other expeditions at the time, if they went well, were at best exercises in survival. Fourteen years earlier, Lieutenant George De Long had penned his last journal entry in Siberia, and nine years earlier Greely had barely escaped alive. But Nansen wrote of shooting stars and "lovely weather." To ease the boredom, he and his partner took long walks in front of their hut. A playful arctic fox amused them and developed a habit of stealing from the camp. Its thefts included, oddly enough, a thermometer. Nansen wrote, "There is furious weather outside, and snow, and it is pleasant to lie here in our warm hut, eating steak, and listening to the wind raging over us." They slept to a point approaching hibernation, to a point at which sleeping became an art. "We carried this art," Nansen wrote, "to a high pitch of perfection, and could sometimes put in as much as 20 hours' sleep in the 24."

❄   ❄   ❄

It is August eighth. I stand in a weed-choked lot just outside Fairbanks, Alaska, one hundred miles south of the Arctic Circle. It is close to sixty degrees. A giant air conditioner drowns out the noise of traffic, wind, and birds. In front of me, built into the side of a hill, is a shed, painted brownish red, a color marketed as redwood but looking entirely unnatural here among the spruce trees. A door leads into the shed and from there into the hillside itself.

This is the permafrost tunnel, built by the U.S. Army Corps of Engineers in the 1960s to test tunneling equipment. The idea was to bore into frozen hillsides, perhaps turning them into missile silos and bunkers. Cold War thinking, so to speak. Now the tunnel is used for research and education. A worn display panel shows visitors a picture of a very calm Dwight D. Eisenhower next to Nikita Khrushchev, whose fist is raised. I am here with a Russian permafrost expert based at the University of Alaska. Pointing out the photographs of Eisenhower and Khrushchev, he laughs at the way the world has changed.

The hill itself is of loess — finely ground particles originating when rocks were pulverized by massive rivers of ice in the Brooks Range, north of here. This grinding goes on today under what is left of the glaciers, but most of it occurred when the glaciers were more extensive, from seventy thousand years ago to a mere ten thousand years ago, and before that, on and off for two and a half million years. The ice, sometimes miles thick, flowed down hills and across valleys, carrying with it stones and boulders and rocks. The bottom of the glacier, with its load of rock, acted as a massive, slow-moving mill, reducing granite mountainsides to dust as fine as flour—the same stuff that today paints the water of glacial lakes azure. But when the glaciers pulled back, the flour was everywhere. Gales blew where warming ground met glacial ice. The gales picked up the flour, scattering it through central Alaska, restacking it in drifts that became hills.

A walk into the permafrost tunnel is a walk through time. The lighting, the air-conditioning system, the signs, the very feel of the place speak of the Cold War. The sweeping scars of the tunneling machine, now decades old, remain frozen in place. And the walls themselves range to more than forty thousand years old. The walls of this tunnel—the earth of this hill—have been frozen solid for forty thousand years.

Frozen soil is not a rarity. Something like one-fifth of the world's land area lies within the permafrost zone. Poke a steel rod into the ground in northern Alaska, and you will hit frozen ground. The same rod will hit frozen ground in northern Russia, northern China, northern Norway, Iceland, and Greenland. It will hit frozen ground on certain mountaintops at the latitude of California. It will hit frozen ground in parts of Patagonia. Late in the summer, the rod will penetrate eighteen inches, thirty inches, three feet, and then hit what feels like bedrock. But it is a bedrock of frozen sand or gravel or fine glacial flour, glued together by ice. In some places, three-quarters of the soil is in fact frozen water. Put a building on this stuff, heat the building and warm up the ground, and the ice will start to melt.

What makes this tunnel unusual is that the government dug into the frozen ground, then kept it frozen. In summer, the massive air conditioner keeps the tunnel chilled near its entrance, where warm drafts sneak past doors. The earth here is like a giant cooler, its outer layers insulating its inside, keeping the tunnel walls in the low twenties. In these latitudes and farther north, the surface expression of frozen ground is visible everywhere, in the form of polygons and frost boils and slumped ground. Here you can walk right through that frozen ground. You get the worm's-eye view.

The place stinks. It is a forty-thousand-year-old smell of mixed mold and musty dirt and cold, something like the smell of a refrigerator that has gone too long unopened. The tunnel is twenty feet in diameter and roughly round in cross section. One passage leads back into the hill more or less horizontally, while another slants downward. We head in horizontally, taking advantage of a metal walkway. Roots stick out from the walls and ceiling. It is easy to imagine that these roots are alive, reaching down from the birch and alder trees growing on the hillside. But in fact we are well below the root zone. Ten feet beneath the surface, the ground never thaws, and living tree roots do not penetrate into permanently frozen ground. The roots in the tunnel walls are the frozen remains of Pleistocene plants. And what is this? The bones of a long-extinct steppe bison: a jawbone, a femur, a vertebra. In the wall, a horn stands frozen in time. The steppe bison was common here thirty thousand years ago but has been extinct for thousands of years.

The Russian tells me of plant material found in the tunnel that was still green after thousands of years. Grass had been covered with snow in a summer blizzard, and then the snow was buried under blowing soil. Likewise, the bison bones had been buried by the blowing loess, preserved for thousands of years. Occasionally, whole animals, flesh intact, are preserved in the permafrost. In 1979, a Fairbanks gold miner found a frozen steppe bison. That is to say, he found not only bones and teeth but a frozen carcass complete with skin and muscles and hair. Claw and tooth marks show that it was killed by an extinct lion. It had frozen so quickly after its death that scavengers could not pick it to pieces. In the holes left behind by the lion's teeth, coagulated blood remained frozen in tiny pools. Shortly after the kill, or during the kill itself, snow may have been falling. The lion may have fed on the carcass for several days or even weeks but abandoned it before spring, leaving meat and flesh and bones behind. One can imagine the lion wandering away, overtaken and bewildered by blowing snow. At first the snow drifts against the carcass. Later, loess carried by wind or a landslide settles on top of the snow. The dead steppe bison is buried. A new layer of permafrost forms. Years pass. Miners dig into the icy ground. A university professor becomes involved. Carbon dating of a piece of skin shows that the bison died thirty-six thousand years ago. The carcass stands today in a glass case at the Museum of the North in Fairbanks, resurrected, looking more like a Texas longhorn than the modern bison of the Great Plains.

We walk past and through different features of frozen ground. I stand beneath an ice wedge — the same sort of ice wedge that forms polygons in the ground farther north. Near Prudhoe Bay, the ground is laced with these things, but they are visible only in their effect on the ground's surface. If the first few feet of soil around Prudhoe Bay were magically removed, the ground would become a honeycomb of ice. The soil, intact, obscures this reality. Here, underground, I can see the wedges themselves. The ancient ground expanded in summer and contracted in autumn, opening cracks. The cracks filled with water, and the water froze. The cycle was repeated again and again. And then the ice wedges were buried under the blowing loess that would become the walls of this tunnel. Looking at an ice wedge in the wall of the tunnel, I see a record of the process. Sediment tracks run up and down its body, marking each year's sequence of cracking and freezing, reminiscent of tree rings. The wedge is more than four feet wide at its top. Conservatively, it took hundreds of years to form.

Water is strange stuff. Most substances, when cooled, contract. This is why thermometers work: mercury shrinks as it cools and expands as it warms. Warmth makes the molecules in a substance move faster. They dance around, bumping into one another. As the temperature increases, they dance faster, and when they bump into one another, they push harder. They need more space. A cooler temperature means slower movement, softer collisions, and less space. This holds true for water, but only to thirty-nine degrees. After that, the water molecules start to line up. The water thickens. Hydrogen atoms in one molecule attract oxygen atoms in others. The process of crystallization begins. At thirty-two degrees, the water starts to freeze. The molecules line up like tiny soldiers in formation, with orderly space between them. Newly frozen water is nine percent bigger than liquid water. Once frozen, if it continues to cool, expansion stops, and like most substances it shrinks.

Small caves and hollows line the tunnel's walls. The ice holding the walls together has not melted, but some of it has sublimated— disappeared into the air as vapor without ever going through a liquid phase. The walls are frozen and steaming at the same time. This is true, too, of snow and ice at the surface—in glaciers, in freezers. The vaporization of ice—evaporation from the solid phase—is the basis of the freezer burn that ruins frozen meat and fish.

Among the ice wedges, veins of ice run horizontally along the tunnel walls like veins of coal in a Virginia mountainside. The Russian calls this "segregation ice." It forms in keeping with another strange property of water: liquid water in finely grained soil is sucked toward colder zones in the soil. We stare at a vein of segregation ice, a one-inch-thick stratum of what looks like almost pure ice.

"People call it cryogenic suction," the Russian says, "as though that explains everything. But cryogenic suction is a very complicated mechanism." In freezing soils, liquid water adheres to soil particles, forming thin layers of water around each grain of soil. Molecules are bound more tightly to thin layers of water than to thick layers of water, and thin layers of water tend to attract molecules of water from nearby thicker layers. When soil starts to freeze, the layers of liquid water turn to ice and become thinner. Liquid water moves from warmer parts of the soil to parts of the soil where ice is forming. Water is sucked toward what is sometimes called the "freezing front." Segregation ice forms.

"I have seen segregated ice one meter thick," the Russian tells me. "I saw it personally." He has heard of segregation ice twenty meters thick—sixty-six feet—but he has not seen it himself. He presents this information as though he does not believe it.

We come to what looks like a small frozen pond or a puddle in the wall of the tunnel. It is shaped like a pond, seen edge on, ten feet across. At the bottom, the ice is dirty, as if it had been filled with silt, and at the top the ice is clear. The Russian calls it an "ice lens." He says that it could have been a frozen pond that was covered by blowing soil and permanently insulated from the summer, but he thinks that it more likely formed underground, a small pocket of flowing groundwater that at some time in the distant past froze into place. We are at this point something like one hundred feet underground, surrounded by soil and ice laid down when steppe bison and mammoths and saber-toothed tigers roamed the surface. Above us, moving toward the surface, the ground is progressively younger. At the surface, the soil is almost brand-new, with four inches or so of fine soil blowing in and settling on top of the hill every hundred years. Below us, deeper underground, in the lower tunnel, it is older.

We wander off the steel grating and into the lower tunnel, heading deeper and farther back into the Pleistocene. Our steps kick up clouds of fine soil. It is strikingly dry, bone-dry. Looking behind us, I see that the air is hazy with the clouds of fine soil. I feel grit in my teeth. I feel it in my hair. It is in my eyes. We look for a moment at what seems to be an old streambed, a layer of sand and stones worked round by flowing water in the distant past. Forty thousand years ago, where we stand now would have been a pretty little stream. Steppe bison would have grazed along its banks. Extinct lions would have stalked the bison and lapped water from the stream. We are forty thousand years underground.

But we are not alone. The Russian once sent a sample from the tunnel's wall to a friend at another laboratory. The sample contained living bacteria. His friend feeds the bacteria and grows it in a laboratory at temperatures hovering around five below zero. This stuff lives and breeds at temperatures where it should be frozen solid.

❅   ❅   ❅

Warmth is not always a good thing. It melts the permafrost. With soil that is as much as three-quarters ice, melting means subsidence. Water flows out of the soil. The ground, melted and drained, sinks. Pools form. The sun warms the pools, setting up convection currents in the water column that pump more heat into the ground. The pools grow wider and deeper. Trees that had once grown on top of the permafrost die in waterlogged conditions. Animal burrows flood. The landscape changes. Build a house on permafrost, and what felt like frozen bedrock beneath the foundation might flow away. The ground might slump. Your house might sag into a water-filled depression. Your neighbors—sourdoughs who know to insulate the ground before they build—might snicker behind your back. Or, this being Alaska, they might laugh in your face. They might invite you to laugh along with them. They might loan you a set of house jacks and offer a deal on a truckload of gravel and cement.

The warmth can come from climate change—a warm winter, a hot summer, a year when the insulating blanket of snow fails to fall from the sky or disappears in early spring. Or the warmth can come from the hyperactive ambitions of human beings. In the 1800s, Alaskan gold miners who had found gold in streams reasoned that there would be more gold belowground, covered by frozen soil. They built fires in mine tunnels to melt the permafrost, loosening gravel from ancient streams buried well below the surface. Later, boilers were used to generate steam at the mine face. Some miners moved away from tunneling and instead stripped away the entire surface, melting the gold-bearing gravel with strings of pipes pumping water into what were called "thawing points." By 1929, a single company had more than ten thousand thawing points operating simultaneously near Fairbanks. Two hundred specialized miners, called "point doctors," pounded the points into the ground and made sure the water flowed. The result: pockmarked ground, thawed and ready for a dredge.

Occasionally, people have built unheated additions onto their homes for storage or as garages, only to see the heated part of the house descend. The walk to the garage is uphill. Certain cabins in the Alaskan bush seem to have been built at odd angles or with deep sags. In Dawson City, northwest of Fairbanks, two frame buildings built during the Klondike gold rush lean together, the ground beneath warmed by the buildings above. To see them is to wonder just how much these people were drinking when they laid the foundations, but they are due not so much to alcohol as to warming and drunkenly subsiding ground.

For victims of hypothermia, rewarming can be fatal. The core temperature of victims continues to drop even after they are brought into a warm environment. Some believe this afterdrop to be nothing more than heat loss from the victim's core to the colder outer layers of the body. Even watermelons, taken from a freezer to a warm kitchen, suffer afterdrop: the warm inner core of the melon loses heat to the colder parts near the rind even as the outer parts begin to warm. Others believe afterdrop occurs when constricted blood vessels near the skin, reacting to the warmer air, reopen, allowing cold blood from the surface to flood the body's core. To make things worse, the cold blood from the surface may be rich in lactic acids, overstressing an already stressed situation. Worst-case scenario: The cold blood hits the heart, causing ventricular fibrillation. The lower heart chamber quivers. The blood stops moving. The victim drops to the ground. The victim dies. This is more common than one might think. It happened after the School Children's Blizzard. It happened to the German soldiers pulled from Norwegian coastal waters in 1940. It happened to sixteen Danish fishermen in 1979. The fishermen, in the water for more than an hour, climbed aboard a rescue

boat, wrapped themselves in blankets, headed to the cabin for coffee, and one by one dropped dead.

And then there is frostbite. Apsley Cherry-Garrard, of Scott's
South Pole expedition, wrote of treating frostbite: "Then you nursed
back your feet and tried to believe you were glad—a frost-bite does
not hurt until it begins to thaw. Later came the blisters, and then
the chunks of dead skin." In Napoleon's campaigns, the men treated
frostbite by warming their frozen extremities next to a fire. As one
cannot actually feel a frozen extremity until the extremity warms
up, it was not uncommon for Napoleon's men to burn themselves
during rewarming. The men learned to rub frozen extremities with
snow. This was painful, too, but it did not cause burns.

During the School Children's Blizzard, rubbing with snow was
the treatment of choice. Addie Knieriem, a young girl who survived
the blizzard, was left with badly frozen feet. As always, the freezing
had started in her skin, between the cells. The ice crystals drew
water from within the cells, dehydrating the cells themselves. Eventually, the fluid remaining within the cells froze. Ice crystals tore
cell membranes. The ice penetrated deeper into her tissues. Blood
vessels froze. Tendons froze. Muscles froze. Her rescuers rubbed
her feet with snow. As her feet warmed, the feeling returned with a
vengeance. She felt as though someone was burning her feet. There
would also have been a terrible, insatiable itchiness. Soon after,
blood cells clotted in her feet. The skin blistered. Her toes turned
black. Her feet began to rot. Her room would have been filled with
the foul stench of gangrene. Her toes were amputated, and then one
of her feet. At this point, with limited painkillers, Addie may have
wondered if warming up had been the wisest choice.

Today frostbitten extremities are rewarmed in warm baths. Painkillers are administered. Antibiotics are used. Rubbing with snow is
discouraged. Amputations remain common. C. Crawford Mechem,
in a 2006 article, says that there are four degrees of frostbite. First-
degree frostbite symptoms include swelling, a waxy look to the skin,

hard white spots, and numbness. By the time third-degree frostbite occurs, symptoms include "blood-filled blisters, which progress to a black eschar over a matter of weeks." In fourth-degree frostbite, there is what Mechem calls "full-thickness damage affecting muscles, tendons, and bone, with resultant tissue loss and sensory deficit." In other words, tissue with fourth-degree frostbite is dead or as good as dead. Addie Knieriem, on the prairie, suffered from fourth-degree frostbite in her feet. On rewarming, Mechem says, one should expect "pain, throbbing, burning, or electric current-like sensations." The last gasp of rewarmed nerve cells comes with intense cries of pain.

Warmth can hurt plants and animals accustomed to the cold. On sunny winter days, pine needles absorb heat from the sun. This may, at first glance, seem like a good thing, a chance for a bit of midwinter photosynthesis. But as the needles warm, the water vapor inside them warms, too. The warmed water vapor leaks out of the needles. The pine tree suffers from what is sometimes called winter desiccation.

And from the world of well-meaning animal conservation: too much warmth in winter kills bats. Indiana bats spend the summer spread out across the eastern United States, but in winter most of them flock to a handful of caves. Half of the world's Indiana bats overwinter in two caves. In 1967, Indiana bats were protected by law because of continuous population decline. What happened? Cave entrances had been modified to prevent cavers from disturbing the bats. The new entrances—now too small for humans but big enough for bats—restricted airflow. Caves breathe: as the temperature changes outside, air moves in or out of the caves. It is not unusual to feel a strong breeze well underground. When the entrances were modified, the caves became asthmatic. Winter temperatures in the caves rose. At Kentucky's Hundred Dome Cave, temperatures rose almost twenty degrees. Hibernating Indiana bats maintain a temperature close to that of the air. At lower temperatures, their hibernating metabolism is very slow. They survive the winter by slowly burning through limited fat reserves. When the temperature increases, their metabolic rate increases. Slowly, during their winter sleep, warm Indiana bats starve to death.

❄   ❄   ❄

It is August fifteenth. It is around sixty degrees. I sit in the rain outside a yurt thirty miles north of my home in Anchorage, Alaska. A yurt is a round, semipermanent tent modeled after those used by nomadic Mongolians. But this yurt is here in Alaska rather than there in Mongolia, and it is furnished with a woodstove, bunk beds, and folding chairs. Beneath me, I can see and hear the Eagle River, which flows from Eagle Glacier, out of sight some fifteen miles upstream. I look periodically for a grizzly sow and her cub, which were seen by hikers just yesterday feeding on salmon in the river. Now the grizzlies are as invisible as a North Slope caterpillar.

The Eagle River valley is U shaped, scraped out by a glacier, without the steep-cut angles of valleys cut by flowing water. Higher up in the mountains around us, other valleys end in waterfalls that stream down the sides of mountains. These hanging valleys mark the surface of the old ice, hundreds of feet up. In the past, the big glacier, the thick one, flowed through this valley, extending up beyond what today are the high mountain passes, and the glaciers in the higher mountain passes intercepted the glacier that flowed through this valley. The glaciers in the high mountain passes flowed into the larger valley glaciers like shallow streams flowing into a deep river, leaving their beds perched well above the bottom of the deep river, but these were streams and rivers of ice, moving with leisurely, mountain-grinding power. Boulders carried by the glaciers litter the valley floor. My son climbs the boulders while I stare at the mountains, imagining this place buried under hundreds of feet of ice. This makes me imagine Manhattan under ice. Fifty thousand years ago, the Wisconsin Glacier overran what would become New York. Somewhat famously, it ground grooves into the rocks at Central Park. What would become New York has been overrun by many glaciers, a complicated coming and going of glaciers that advanced and retreated over thousands of years, a cosmopolitan mingling of ice and the effects of ice that make the modern world seem inconsequential. The glaciers in New York are gone now, as is most of the glacier here in the Eagle River valley. All that is left is a pathetic remnant near the head of the valley. The world has warmed. In national parks, there are signposts marking the extent of certain glaciers in 1959, in 1965, in 1970, and so on to the present day. Trees grow around these signs. The glacier itself might not be visible until you are standing next to a sign that says 1959. In ten years, it might not be visible from beyond a sign dated 1970. In twenty years, it might be gone altogether. People say the glaciers in Alaska have retreated. In fact, they have melted, withered, and in some cases disappeared, warmed and thawed into miniatures of their former selves or warmed further until nothing is left but a ghostly memory running over ground rock.

Tired of the rain, I wander back into the yurt. The park management has left a guest book. Last January, someone wrote, "When we got here yesterday it was three degrees out and not much warmer inside. We got the stove going and got it to seventy inside."

Last February, someone wrote, "I hope that the next person that stays in this yurt will have a better experience than we did. P.S. It was minus fourteen Fahrenheit when we got in. We brought my dog with us. Her name is Biscuit."

Also in February, someone wrote, "The worst part however had to be the outhouse so cold on the bum."

I write nothing. Instead, I stoke the fire in the woodstove. I listen to the rain hitting the tent, audible over the river's rushing. The next morning, although it is August in a warming climate, there is new snow on the tops of the mountains. It is termination dust, early this year, a dusting of snow marking the end of summer.

❄   ❄   ❄

On the prairies, hypothermic and frostbitten children were carried into tiny houses. When times were good, these houses might have been heated with coal or wood. Some families burned dried buffalo bones. Some, when they had to, burned what they called poor man's coal or prairie coal—little bundles of hay, manually twisted together and fed into a hay burner. The trouble with prairie coal was that it burned quickly and did not put out as much heat as wood or coal or bones. Hay burners had to be fed almost constantly, which meant that someone was almost constantly twisting together handfuls of hay.

Richard Byrd, in his 1933 solo adventure in Antarctica, would strip off his mask and diving-suit apparel inside his hut in early May—autumn on the southern continent. He wrote of "the small sounds of the hut": ticking clocks, chattering instruments, and, importantly, the hiss of his oil-burning stove. The stove, unbeknownst to Byrd, was leaking carbon monoxide, slowly poisoning him. It left him weak and confused. "I was at least three hours getting fuel," he wrote, "heating the engine, sweating it into the shack and out, and completing the other preparations. I moved feebly like a very old man. Once I leaned against the tunnel wall, too far gone to push the engine another inch. You're mad, I whispered to myself. It would be better to stay in the bunk and cut out paper dolls than keep up this damnable nonsense." He became despondent. He ignored an overturned chair. He could not bring himself to read his books. He could not get warm, inside or outside. "What baffles me," he wrote, "is that I have no reserve strength whatever." Eventually, he radioed his support team. He did not want anyone to know how his condition had deteriorated, so he asked a question, by his own accounting, in "an offhand manner." His request was simple: "Have Dr. Poulter consult with the Bureau of Standards in Washington and find out: (1) whether the wick lantern gave off less fumes than the pressure lantern; and (2) whether moisture in the kerosene or Stoddard solvent (in consequence of thawing rime in the stovepipe) would be apt to cause carbon monoxide." A few days later, he had his answer. Yes, carbon monoxide might be a problem. Poulter, however, did not suggest any cures that Byrd had not already tried. On August 11, 1934, near the end of the southern winter, help arrived. Byrd was, by this point, too weak to continue on his own. Without outside help, his heater would have killed him. He would have succumbed to the fumes.

The human furnace, for a typical adult male, burns through something like seventeen hundred calories a day just to get by. When the body is cold, the burn rate can go up another four hundred calories or so just to stay warm. Shivering can increase heat production four times, but only until the body's supply of glycogen — a form of sugar stored in the liver and easily converted to the glucose needed by active muscles — is gone. Then shivering stops, and the body temperature plummets.

At rest, the organs generate most of the body's heat. The brain by itself accounts for something like one-sixth of the total. In motion, the muscles take over. Moving uphill on skis can burn through more than a thousand calories an hour. Two hours on skis can burn more calories than a full day at rest. Physically fit men, such as polar explorers, can maintain activity levels that burn more than six hundred calories an hour all day long. It is difficult, at this level of activity, to feel satiated. "A pemmican soup," wrote Frederick Cook of a meal he enjoyed close to the North Pole in 1908, "flavored with musk ox tenderloins, steaming with heat — a luxury seldom enjoyed in our camps — next went down with warming, satisfying gulps. This was followed by a few strips of frozen fresh meat, then by a block of pemmican. Later, a few squares of musk ox suet gave the taste of sweets to round up our meal. Last of all, three cups of tea spread the chronic stomach-folds, after which we reveled in the sense of fullness of the best meal of many weeks."

Through most of the nineteenth century, the explorers ate pemmican. Pemmican—the real thing, as opposed to the beef jerky marketed as pemmican today—was dried and pulverized meat and bones and berries. Think of something approximating perhaps dried Spam with berries. The meat and bones often came from moose or bison or elk. It tasted as good as it sounds. George Tyson, afloat on ice in 1873 with Eskimo hunters and their wives, described a meal: "We pound the bread fine, then take brackish ice, or saltwater ice, and melt it in a tin pemmican can over the lamp; then put in the pounded bread and pemmican, and, when all is warm, call it 'tea,' and drink it. It reminds me very much of greasy dish-water."

When possible, explorers supplemented their diets with fresh game, in part for the sheer pleasure of fresh food and in part because of the knowledge that it kept scurvy at bay. There might be seal meat or caribou or fox. Polar bear was not an unusual meal. Many men ate their meat frozen, a habit they learned from the Eskimos. It was noted more than once that food could be cooked with fire or with ice. Father Henry, living in his ice cellar in Canada's Northwest Territories, subsisted on frozen fish. "For six years," wrote a journalist who knew Father Henry,

he had been living on nothing but frozen fish, and he was none the worse off for it. When he awoke he groped on the ground, picked up a great chunk of fish frozen so hard that he had to thaw it out a little with his lips and breath before he could bite into it, and with this he regaled himself.... Boiled rice warmed you while you ate it, but its warmth died out of you almost as soon as it was eaten. Frozen fish worked the other way: you did not feel its radiation immediately; but twenty minutes later it began to warm you and it kept you warm for hours.

All of the warm-blooded animals need food to stay warm. Some animals conserve calories by hibernating at cooler than normal body temperatures, but others do not. Polar bears, other than pregnant females, hunt seals through the year. Arctic foxes roam the sea ice in winter. Beneath the snow, subnivean life churns on. Lemmings, the size of gerbils, dash through tunnels of hoarfrost at the bottom of the snowpack. Smaller animals lose heat more quickly than larger animals. For their size, smaller animals have more exposed skin— more surface area to cool off for every ounce of fat, muscle, and brain tissue. For their size, they need to burn more calories than larger animals, yet they are not big enough to store much fat. In colder climates, a small animal that cannot store fat cannot hibernate. The smallest hibernator in the far north is the ground squirrel, several times larger than the lemming. And so lemmings eat through the winter. They gnaw on twigs and branches under the snow. Occasionally, when times are hard, they eat each other.

Farther south, in the American suburbs and frosty cornfields and icy parks, birds foolish enough to stick around through the winter eat to stay warm. It would not be entirely wrong to say that they feed desperately. Golden-crowned kinglets, birds not much larger than a grown man's thumb, overwinter in places such as Vermont and Maine. Despite thick feathers, they lose heat quickly. Winter ecologist Bernd Heinrich, curious about what they ate, killed a few one winter. "I shot the first kinglet at dusk when the bird's stomach would presumably be full," he later wrote. He "took its body temperature as soon as it hit the ground." The little winged furnace was thermoregulating at 111 degrees. Its stomach was full of tiny caterpillars, previously thought to overwinter as pupae. "To find out how quickly a fully feathered kinglet loses body heat," Heinrich wrote, "I experimentally heated a dead kinglet and then measured its cooling rate." At thirty below, the tiny dead bird lost more than five degrees every minute. Alive, at this temperature, the bird would have had to forage almost nonstop. Heinrich followed four of the birds around on a January night. With windchill, the temperature was fourteen below. "The birds foraged tirelessly, without pause," he wrote. "I timed them at an average of 45 hop-flights per minute, without any apparent change of pace."

Cold, really, is like malaria. If it does not kill you, it will help you lose weight.

❄   ❄   ❄

It is August twenty-first and fifty degrees. I am wandering along the length of Point Brower, just east of the Prudhoe Bay oil field, with two botanists, a father-son team. We step on diminutive willows and dryas and saxifrage. We splash through puddles full of cotton grass. We peer into shallow freshwater pools full of tiny copepods. There are also crustaceans called tadpole shrimp—not shrimp at all, but from a group called branchiopods, distant cousins of water fleas and closer cousins of brine shrimp, including the ones marketed as sea monkeys. Soon all of these plants, all of these puddles, and all of these little pools will freeze solid. Everything in them will turn to ice. All of this life will be suspended.

I ask my comrades to watch for caterpillars. Though skunked so far, I have not given up. A lemming scurries across the ground in front of me like a tiny furry pig on espresso. It dashes along minia-ture pig trails between clumps of plants and through tunnels under the leaves. Neither tunnels nor trails are more than two inches wide. Soon enough, the trails and tunnels will be buried under snow. The lemmings will grow their winter digging claws. They will go about their business under the snow, in their icy white grottoes, chewing on frozen willow stems. The snow will protect them from the bitter cold and wind of the surface.

At the end of Point Brower, which juts out into the Beaufort Sea east of Prudhoe Bay, sit the remains of three human homes. They were sod huts more than homes, little rectangular boxes ten feet on a side, before their roofs collapsed. The oldest of them, possibly built more than two centuries ago, when Russia ran Alaska as a frontier province, is now nothing more than a low mound of sod. The second oldest has been used as a trash heap and is full of rusted cans and broken glass. The youngest—the remains of a roof still in place, its sod walls full of lemming tunnels and littered with piles of their droppings—could be marketed as an Arctic fixer-upper, beachfront, ocean views, with kerosene tins and rusting cans that suggest it was used as recently as fifty years ago.

This is treeless country. Structural members in sod homes like these might be bones from bowhead whales or stout driftwood tree trunks. It is Canadian driftwood, carried down the Mackenzie River in Canada to float along the Beaufort Sea before washing up on these gravel beaches, a drift of more than two hundred miles. There are, too, scattered boulders on these beaches, carried here on Pleistocene ice sheets. The sod homes, in modern times, were heated by bottled natural gas or kerosene, and before that by driftwood and seal oil. In some areas, the locals once burned dried tundra sod soaked in natural oil seeps, where crude oil finds its way to the surface from underlying deposits. The place, for all its wildness, has a feeling of history unusual to find in Alaska, a feeling that the people who came before left a mark, that what might look at first like an untamed coast has been home to people for generations, and that these people traded with other people, and built homes and fostered dreams and ambitions of their own through long, cold winter nights and breezy sunny summer days on the shore of an icy sea that to them did not seem at all untamed.

One of the botanists asks me about a bird. "It's a long-billed dowitcher," I tell him. Soon it will fly off to overwinter in California or Mexico or as far east as Louisiana or Florida, like some of the Prudhoe Bay oil field workers. This time of year, when the plants are still green, the Arctic seems unnaturally quiet. Over the past few weeks, most of the birds have left. There are some stragglers, though, like the dowitcher. The geese and the larger ducks have not yet left. And tiny Lapland longspurs still flit around happily. One, foraging on the ground near me, almost walks over my boot. Soon they will head south to overwinter in suburban parking lots and subdivisions, happily taking handouts from bird feeders through the winter.

The same botanist who asked about the bird finds what has eluded me—a caterpillar, and then, almost immediately, another. The caterpillars are mostly deep brown, the color of chocolate, but with black trim. Light gray hairs cover their bodies. The hairs look like bristles, but they are surprisingly soft. The hairs slow down airflow, trapping warmth and moisture in a boundary layer around the beast's body. I let the caterpillars crawl around on my hand. They will make wonderful pets. I will store them in the freezer while I travel. And, if this does not work out, I suppose I can eat them, as Greely's men did. I name one of them Fram, after the boat that Fridtjof Nansen intentionally froze into the ice. I name the other one Bedford, after James Bedford, the retired psychology professor who, immediately after his death from kidney cancer, had his body immersed in liquid nitrogen. Bedford—the person, not the caterpillar—remains frozen in a facility in sunny Arizona. Although I cannot actually tell Fram the caterpillar from Bedford the caterpillar, neither can anyone else. I empty my lunch bag into my knapsack, throw a few willow sprigs in the empty bag, and then bag the caterpillars themselves.

SEPTEMBER

It is September fifth. On the North Slope, the temperature is thirty-nine degrees above zero. On the North Pole, it is just below freezing and overcast. On the South Pole, it is sixty-five below. In Vostok, sitting at 11,484 feet of elevation in Antarctica's Russian sector, the thermometer reads ninety-seven below. I am in none of these places. Here, in Windsor, England, it is seventy-three degrees and too hot to stand in the sun. I stand instead in the shadow of Windsor Castle. Rumor has it that the queen is present. Security is openly active. My request for an interview has been flatly refused.

Windsor Castle has well over a hundred rooms. They have names such as the Green Drawing Room, the Crimson Drawing Room, and the Octagon Dining Room. Ceilings tend to be high. The outer walls are stone. Some of the windows are narrow slits, just wide enough to suit an archer, but others are much larger, well built to let in the cold of British winters. The doors could hardly have been better designed to lose heat. The round towers were never meant to hold the weather at bay.

William the Conqueror chose the site nine hundred years ago. Over the centuries, it has been enlarged and renovated. Parts of the castle originally built of wood are now stone.

It is a shame that my interview was refused, as I had but two questions for the queen. "How much is your heating bill?" I hoped to ask her. And, for follow-up, a delicate question about the quality and use of royal long underwear in winter within the stone walls of a drafty nine-hundred-year-old castle. Silk, I would guess, but verbal confirmation from the queen is needed.

What I really want to know is this: How did the castle's occupants do during the Little Ice Age, starting perhaps as early as the fourteenth century and running until around 1850? And how did they do when Mount Tambora blew its volcanic top and ushered in the Year Without Summer? One can imagine the king's voice echoing down a long stone hallway: "Break out the royal long johns, Squire. It looks to be a cold one again."

❄   ❄   ❄

On April 11, 1815, Mount Tambora, on an island called Sumbawa in Indonesia, exploded. This was no ordinary eruption. Four thousand feet of mountain summit disappeared during three months of tremors, rumblings, and lava and ash eviscerations. Twelve thousand people on Sumbawa died. More than forty thousand on the neighboring island of Lombok starved when their ash-covered crops failed. A British resident of Java, more than two hundred miles from the blast, wrote, "The atmosphere appeared to be loaded with a thick vapour: the Sun was rarely visible, and only at short intervals appearing very obscurely behind a semitransparent substance." Sir Thomas Raffles, then the British lieutenant governor of Java, wrote of violent winds carrying away men, horses, and cattle. The volcano discharged a hundred times more ash than was discharged by the 1980 eruption of Mount St. Helens in Washington State. It dumped more dust than Krakatau in 1883. And, to make things worse, it was the third major eruption since 1812. Soufrière, on St. Vincent Island in the Caribbean, had blown in 1812, and Mount Mayon, in the Philippines, had gone in 1814. Volcanic dust choked the stratosphere.

Dust in the stratosphere acts like a translucent shade on a window. It blocks the sun. This in itself is enough to cool the earth, but it gets worse. Decreased warmth from the sun changes wind and current movements in the Northern Hemisphere. Cold Arctic air moves south. Europeans and Americans called the year after the Mount Tambora eruption the Year Without Summer or the Poverty Year. The laconic farmers of New England referred to it simply as "Eighteen Hundred and Froze to Death." There were novelties such as flesh-colored snow in Hungary, red and yellow snow in Italy, and blue and red snow in the eastern United States, all the result of ash captured in snow clouds. An English vicar wrote, "During the entire season the sun rose each morning as though in a cloud of smoke, red and rayless, shedding little light or warmth and setting at night behind a thick cloud of vapor, leaving hardly a trace of its having passed over the face of the earth." Summer temperatures were as much as eight degrees colder than normal. Violent thunderstorms with hail were unusually common. By the middle of summer, people were worried about crops. On the twentieth of July, the *Times* of London reported, "Should the present wet weather continue, the corn will inevitably be laid and the effects of such a calamity and at such a time cannot be otherwise than ruinous to the farmers, and even to the people at large." During this period, a typical family could spend two-thirds of its income on food. In the United States, rivers as far south as Pennsylvania carried ice in July. It snowed in New England in June. And there were beautiful sunsets through the veil of stratospheric dust.

Mary Shelley was holed up in Lord Byron's lakeside retreat near Geneva in the summer of 1816. The weather kept Byron's guests indoors, and he challenged them to come up with ghost stories. Shelley came up with *Frankenstein,* which was published two years later. The popular impression of the novel today is based on movies that share only the name and a monster with the book, but the novel starts with letters from an Arctic explorer. The explorer spots a dog-sled pulling a strange creature, the living thing mysteriously created by Dr. Frankenstein. Writing from an icebound boat, the explorer soon saves Frankenstein himself from the ice. Frankenstein tells the story of his creation, of how it murdered his wife, and of his obsession with tracking and killing the creature. Frankenstein dies on the boat. The creature boards the boat and looks over its dead creator. A conversation ensues between the explorer and the creature, running a few pages. The creature, saying farewell not only to the explorer but to all mankind, leaps through a cabin window, landing on an ice floe, and drifts off into the Arctic night.

While Shelley dreamt this up in the comfort of Byron's home, the people outside suffered increasing hardships. Grain and potato prices tripled. More than thirty thousand Swiss were without jobs. They ate sorrel, a weedy vegetable then considered most fit for horses. They also dined on a form of lichen and, when they could get them, cats. The following year became known as the Year of the Beggars. In the United States, New England farms were wiped out. Thousands migrated westward toward richer soil. Among the migrants was Joseph Smith, who would later found the Church of Jesus Christ of Latter-day Saints. In Britain and France, the cold weather led to food riots. Whereas wealthier individuals may have been able to feed themselves, the scarcity of oats made it increasingly expensive to feed horses. This made alternative forms of transportation attractive, leading Baron Karl Drais to invent the Draisine, also called the velocipede or *Laufmaschine* (running machine), but best described as a steerable wooden scooter. The Draisine would eventually evolve into the bicycle.

The Year Without Summer was a harsh year during a harsh set of centuries. Centuries earlier, ending sometime around the fourteenth century, Europe had enjoyed temperate weather. England and France came of age during what has since been called the Medieval

Warm Period, stretching from about 800 to around 1300. Vineyards thrived in a warmer England. Although most people depended on subsistence farming methods comparable to those used in the worst of today's developing world, populations grew. With good weather, subsistence farming provided basic necessities as well as a surplus, and the surplus supported cathedral building, monks, and the development of trade. This was the period when Norse colonists settled Iceland and Greenland. During colonization, the shores of Iceland were often ice-free, and parts of Greenland's coastal zone were as green as its name implies.

But something changed. By 1300, reaching the Norse colonies meant sailing far offshore to avoid ice. The Norse settlements in Greenland became less green and were abandoned. In 1492, Pope Alexander VI wrote a letter about commerce to Iceland, saying "shipping to the country is very infrequent because of the extensive freezing of the waters—no ship having put into shore, it is believed, for eighty years." Mountain glaciers expanded in Scandinavia, Alaska, China, the Andes, and New Zealand, with permanent mountain snow and ice occurring more than three hundred feet lower than it had just a few centuries earlier. Some European glaciers reportedly advanced hundreds of feet each month, even in summer. Loss of grazing land and crop failures from shortened growing seasons translated quickly into hardship, but at the same time, certain fish species moved south. "The herrings," wrote British geographer William Camden in 1588, "which in the times of our grandfathers swarmed only about Norway, now in our times...swim in great shoals round our coasts every year." In 1610, John Taylor of central Scotland wrote, "The oldest man alive never saw but snow on the tops of divers of these hills, both in summer as well as in winter." The Thames froze repeatedly, providing an icy thoroughfare through central London. Diarist John Evelyn, in an entry dated January 24, 1684, wrote, "Frost...more & more severe, the Thames before London was planted with bothes [booths] in formal streets, as in a Citty...the trees not onely splitting as if lightning-strock, but Men & Cattel perishing in divers places, and the very seas so locked up with yce, that no vessells could stirr out, or come in." In 1692, a French official wrote, "The poor people are obliged to use their oats to make bread. This winter they will have to live on oats, barley, peas, and other vegetables." Between 1695 and 1728, Eskimo kayakers, apparently hunting along the southern edge of the polar ice and then perhaps blown off course or intentionally exploring south of their normal hunting grounds, were spotted off Scotland's Orkney Islands. In at least one case, kayaks were seen as far south as the River Don near Aberdeen.

This was all part of what has come to be called the Little Ice Age. François Matthes, a glacial geologist, first used the phrase "little ice age" in 1939: "We are living in an epoch of renewed but moderate glaciation — a 'little ice age' that already has lasted about 4,000 years." Later, the Medieval Warm Period was recognized, separating a cold snap that had started around 2000 B.C. from the cold snap that started sometime around the fourteenth century. But it is not right to think of these periods as cold snaps. They were, on average, colder than earlier times, but only by a couple of degrees. Summers could be quite hot, but winters were colder and longer than at other times, bringing the average down.

Dust from volcanoes played a role, but only well after cooling had started. A disruption in warm ocean currents may have played a role, too, but then one would have to ask what disrupted the currents themselves. It is known that changes in the sun of only a few tenths of a percent can change the earth's climate. In 1711, the English astronomer William Derham commented on "great intervals" with no sunspots between 1660 and 1684, at a time when stargazing was increasingly popular: "Spots could hardly escape the sight of so many Observers of the Sun, as were then perpetually peeping upon him with their Telescopes... all the world over." Although much has been made of this, no one has explained why sunspot activity decreased or exactly how this might explain climate change.

Snow played a role in the making of the Little Ice Age. Snow is an almost perfect reflector, sending heat and light back into space with remarkable efficiency. When snow covers the ground, the earth does not warm as quickly as it does when the snow is gone. A snow-covered planet is a cold planet. But this leaves the question of cause unanswered, since the cold would have had to come before the snow, at least initially.

Whatever the cause of the Little Ice Age, and as significant as it may have been to the people it iced, it was nothing compared to what is normally thought of as the Pleistocene Ice Age, the series of cold snaps that led to massive glaciation. Despite its name, the Pleistocene Ice Age started during the late Pliocene, some two and a half million years ago, with the spread of ice sheets in the Northern Hemisphere. For two and a half million years, it waxed and waned in intensity. The ice sheets advanced and retreated, marking the world's timeline with glacial and interglacial periods. The ice sheets expanded during glacials and retreated during interglacials. The Pleistocene Ice Age is only now appearing to peter out. The last glacial ended ten thousand years ago. Today's interglacial leaves us with nothing more than Antarctica, scattered mountain glaciers, and the remnant ice sheet of Greenland. The Pleistocene Ice Age gave the world woolly mammoths, the ice-carved valleys of Alaska, and the beauty of Scotland. But for all this, the Pleistocene Ice Age was only one of at least four major ice ages in the world's timeline. And it was no more than a chilly breeze compared to the ice age of seven hundred million years ago, when the entire planet may have been more or less frozen, a godforsaken snowball hurtling through space.

❄ ❄ ❄

It is September tenth and sixty-eight degrees on the lower slopes of Ben Nevis, Scotland's highest peak. Scotland's reputation for mountains comes from their abundance, not their height. Ben Nevis is a mere 4,409 feet tall, a bump trying to become a hill and calling itself a mountain. Nevertheless, the tourist information office issues dire warnings. Hikers should have warm clothes. They should be prepared to overnight in cold, wet conditions. The mountain creates its own weather. Snow can come at any time. Fog can cloak the route without warning. Hikers should have adequate food. They should have an Ordnance Survey topographical map and a compass, and they should know how to use both. In short, they should take the hike seriously.

My companion and I walk the lower slopes in T-shirts and shorts. Between us, we have two scones left over from breakfast. We have the map from our walking guide to Scotland, a small, unreadable sketch that shows a zigzagged switchback trail leading toward a summit. My compass is in Alaska. Her compass is in Holland. In short, we have no choice but to climb the mountain.

It is an upward slog, steady and constant. We round a bend, and wind slaps us head-on. We are both sweating, and the cold wind is a comfort. We pick up the pace.

Scotland is a landscape formed by ice. To drive up from London is to drive into glaciated terrain. South and west of the extent of glaciation, it is flat and neatly cultivated. Then it becomes the rolling hills of glacial moraine. Here the glaciers coming from the north slowed, retreated, advanced again, finally stopped and pulled back. With each hesitation, millions of pounds of crushed rock and dirt were dumped, leaving moraines behind to form the rolling hills. Farther north, small steep mounds called drumlins pock scattered valleys. Drumlins formed when hard rock under the glacier slowed down its forward movement. Tripped by the obstruction, the glacier dumped part of its load of rock and dirt. Most of it piled up just upstream of the obstruction—up-glacier, toward the source of the creeping snow and ice—forming a blunt slope, while the downstream tail tapered to ground level. Drumlins in Scotland often occur in bunches that are sometimes called "baskets of eggs."

And that is what they look like: baskets of earthen eggs on valley floors. And with the drumlins, scores of erratics — boulders carried by glaciers — pepper the valley floors and the shallower slopes of the mountains. The erratics — odd shapes, some small enough to be lifted by one man and others the size of a van — stand or lie on their sides. Sheep wander around them, grazing, chewing stupidly in the midst of grandeur.

Farther north, in the Scottish Highlands, the mountains and valleys have been shaped by glaciers. It could almost be Alaska. The ridges are often knife-edged, sharpened by glacial erosion into what some call arêtes. Arêtes form when glaciers freeze around rocks. Glacial water finds its way into cracks and crevices, refreezes, expands, shatters the rocks, and then sweeps the broken mess downhill. A hollow called a corrie or a cirque is left in the side of the mountain, under the knife-edged peaks, with the soft curve of a giant easy chair. The mountains become steep sided, the valleys U shaped, the ground scraped and abraded and scarred by grinding rocks and ice. The brutality of ice leaves mountains brutally beautiful.

We stop at the top of Ben Nevis to eat our scones. Here, though only a few thousand feet above our starting point, the temperature is in the fifties. With windchill, it is in the low forties. Our clothes are damp. We cool quickly, sitting behind what is left of a stone building. The roof is gone, and the upper walls have collapsed. Trash litters the interior. Tourism is part of the mountain's history and has been for some time. At one point, two women offered a bed-and-breakfast on top of the mountain — ten shillings for lodging with breakfast, three shillings for lunch. Before that, the mountain hosted a weather station. In the late 1800s, with a climate still hungover from the Little Ice Age, workers reported having to tunnel thirty feet through snow to get in and out of the station. On some days, the wind could knock over a grown man, and the workers moved about roped together. The average temperature was just above freezing. On cold days,

Ben Nevis touched the zero mark. Average annual precipitation was more than thirteen feet. The weather station closed, perhaps wisely, in 1904. Today its remains stand beneath a metal-doored survival shelter. I peek inside to find an otherwise bare cell littered with trash, like the stone-walled ruins outside.

My companion, who suffers from Raynaud's disease, shows me her hands. As soon as she stopped walking, she cooled, and the blood vessels in her fingers constricted. The fingertips have taken on the color of a yellowing corpse. The lower halves of her fingers and the palms of her hands are a faintly mottled mauve, a most unnatural color for human flesh. At this point, she cannot work a zipper. I touch her fingers. They are cold. She tells me they feel numb. I tell her they feel dead. When they warm, they will sting. The disease affects something like one in twenty people. It comes in varying shades of severity. Women are more susceptible than men. Raynaud's is best prevented, physicians say, by staying out of the cold. It is best treated by rewarming. The disease is more of an annoyance than a serious threat. When I say this, I mean an annoyance for her. For me it is a curiosity. As we move down the mountain, I entertain myself by stopping intermittently to observe her recovery. At one point, her fingers are striped with mauve and pale yellow bands. Sadly, I am not carrying a camera. It occurs to me that Raynaud's would be deadly if it prevented someone from striking a match to start a fire or tying a bootlace or cinching down the harness on a dogsled. Raynaud's would have had no place on Greely's crew, or Scott's, or Shackleton's.

The glacial ice is long gone from Scotland, absent now for thousands of years. But ice still sculpts the landscape. In winter, pipkrakes churn the soil. Pipkrakes give the collapsing crunch to frozen ground in early winter. They are known by other names: needle ice and mush frost, *Kammeis* to Germans, *shimobashira* to Japanese. They grow within the soil, from the bottom up, when the temperature drops below freezing. Water in the soil freezes, and the freezing water sucks more water up from beneath. Vertical crystals grow. They grow as much as half an inch in a single night. As they grow, they push the soil upward. The churning of the ground pushes the fine-grained soil to the surface, where it is blown away in areas with high winds, leaving behind shallow hollows and causing what some farmers call soil deflation. If you are in the business of growing crops, soil deflation deflates your assets. Certain farmers, unable to bear the thought of blowing assets, compact the ground where pipkrakes form.

Here in the Scottish Highlands, pipkrakes have sculpted steps into the slopes. These steps are anything but subtle. Pipkrakes lift the smaller grains of soil away from larger pebbles and rocks. At a slower pace, pipkrakes lift the pebbles away from the rocks, and at an even slower pace, they lift the smaller rocks away from the bigger rocks. In Canada, pipkrakes have lifted rocks weighing upward of one ton. On steep slopes with scattered vegetation, the fine soil pushed to the top blows away. The pebbles tumble downslope. The rocks settle back into place. All of this occurs at differing paces. The end result can be solifluction, the slow and irregular sliding of a hillside. Under the right conditions of soils and slopes and moisture, solifluction leaves what looks like a regular pattern of cryoturbation steps. These steps sometimes form diamond patterns on the slopes, but they also form long parallel terraces, each progressively lower platform a step away from and a step below the next. It appears as if some bored druid or some Iron Age tribe had carved steps up the sides of these mountains.

Moving downward, we take a shortcut across scree, loose rock and gravel splintered by ice. On scree, it is difficult to remain upright. My companion, her hands still numb from the Raynaud's, is afraid of falling, rightfully scared because her numb hands will be little help in breaking the fall. I scurry ahead. Below the scree, I find a hollow of deflated soil that shaves a notch off the wind and nap while waiting for her to catch up.

※   ※   ※

Recognition of the Pleistocene Ice Age is a surprisingly recent phenomenon. Two hundred years ago, no one would have suggested that ice sheets many miles thick had once covered most of Britain, most of Norway and Switzerland, New York City and Boston. People knew of ice sheets in Greenland, and they were learning about the ice of the far north. James Cook and others had sailed far enough south to have been stopped by Antarctic ice. They knew of alpine glaciers, and they knew of the huge boulders scattered around the lowlands—the erratics sitting on top of soil, with no hint of how they might have gotten there, miles from the nearest source of this sort of rock, far too big to carry. There was some thought that the boulders had been carried there by ice, but floating ice during the biblical Flood rather than vast ice sheets hundreds or thousands of feet thick and stretching hundreds of miles to the north. As for other signs of glaciation—the strange hills of soil that would become known as moraines, the baskets of eggs called drumlins, and the odd scoring often found across the faces of exposed bedrock—they were easy enough to overlook and ignore.

Charles Darwin, after learning about ice ages, chastised himself and others for missing the obvious. "I had a striking instance how easy it is to overlook phenomena," he wrote, "however conspicuous, before they have been observed by anyone.... Neither of us saw a trace of the wonderful glacial phenomena all around us; we did not notice the plainly scored rocks, the perched boulders, the lateral and terminal moraines."

The observer who pointed out the obvious was Louis Agassiz, a Swiss scientist. As a student in Munich, he cataloged fish from the

Amazon for one of his professors. This did not mean traveling to the Amazon itself. Agassiz traveled instead through pickled samples, sketching whole fish and scales and bones. He explored the vistas of a laboratory bench and named the fish that he encountered in those wanderings. In 1829, while still a student and despite never having been to Brazil, he published *Brazilian Fishes,* making his first mark as a naturalist. Ten years later, by then widely recognized for his work, Agassiz vacationed near Bex in the Swiss Alps. There the geologist Jean de Charpentier ran a salt mining operation. Charpentier, in part through his association with an engineer named Ignatz Venetz, was convinced that the Swiss glaciers had shrunk over time. He took Agassiz to look at boulders along the faces of existing glaciers, then showed him others farther downhill, dumped there sometime in the past. They were of a kind of rock that might not occur for miles around, and as often as not they were stuck precariously on valley walls or left standing in odd positions. They were erratics. The two men also saw curved beds of gravel and earth — moraines — along the faces of glaciers that matched those farther downslope, and they saw grooves and scratches cut into bedrock.

A year later, in 1837, Agassiz presided over a meeting of the Natural History Society of Switzerland. In his introductory speech, when he was expected to talk about fossil fish, he sprang the idea of an ice age. Although Charpentier knew that the alpine glaciers had once covered more of the Alps than they currently did, Agassiz went further. He described a sheet of ice extending from the North Pole to the Mediterranean. He knew that some would view this as harebrained. "I am afraid," he said, "that this approach will not be accepted by a great number of our geologists, who have well-established opinions on this subject, and the fate of this question will be that of all those that contradict traditional ideas."

Three years went by before Agassiz published *Études sur les Glaciers.* "In my opinion," he wrote, "the only way to account for all these facts and relate them to known geological phenomena is to assume that...the Earth was covered by a huge ice sheet that buried the Siberian mammoths and reached just as far south as did the phenomenon of erratic boulders." He was wrong about many things. "The development of these huge ice sheets must have led to the destruction of all organic life at the Earth's surface," he wrote. "The land of Europe, previously covered with tropical vegetation and inhabited by herds of great elephants, enormous hippopotami, and gigantic carnivore, was suddenly buried under a vast expanse of ice, covering plains, lakes, seas, and plateaus alike." In fact, the Ice Age was not sudden, it did not bury all of Europe under ice, and it did not destroy all organic life, but his general premise was correct. Made famous by this premise, he moved to the United States to take a professorship at Harvard. He died in 1873 and was buried in Cambridge, Massachusetts, under a granite erratic shipped from a moraine in Switzerland. Its transport across the Atlantic made it the most erratic of erratics. By that time — nearly six decades after the Year Without Summer and Shelley's *Frankenstein,* a few years before Greely and a handful of his men barely survived their experience in the Arctic, and fifteen years before the School Children's Blizzard — the existence of a great ice age was considered a fact. The world knew of the Pleistocene glaciations.

Speculations about the cause of the Ice Age abounded. James Croll, a self-educated Scotsman who had run a tea shop and worked as a millwright before becoming known for his scientific contributions, wrote, "We may describe, arrange, and classify the effects as we may, but without a knowledge of the laws of the agent we can have no rational unity." Some thought that the sun might change over time. Others wondered if the earth might occasionally drift through cold regions of space. Some thought that the earth might have somehow rolled, so that the poles were at the equators. But Croll built his work on a proposal put forward in Charles Lyell's famous *Principles of Geology* in 1830. Lyell had skeptically suggested that someone should look into the possible influence of astronomical conditions. Someone, Lyell had suggested, should do the math. Croll decided that he was the someone for the job.

This was in a time before computers, but it was known that the earth's orbit was elliptical and that the ellipse changed over time. As other planets tugged at the earth, its orbit could become more round or more stretched. When its orbit was stretched, the earth would be farther from the sun during the winter and might receive less light. It was also known that the earth's axis was tilted; the earth leaned at an angle to the sun, making winter days shorter than summer days. More important for Croll, it was known that the earth's axis wobbled over time. And Croll knew that periods of extensive glaciation during the Ice Age had in fact come and gone and come and gone through repeated cycles, that the Ice Age was not a single event but rather a pattern of events that had to be explained. Building on the work of others, he looked for a cause that could explain these cycles of relative cold and relative warmth, or more correctly, relative cold and relatively less cold. He saw that the combination of changes in the earth's orbit and a wobbling axis would lead to at least mild changes in the heat coming into the Northern Hemisphere. He realized that slight cooling could mean more snow. He knew snow to be a nearly perfect reflector of heat. On a cold, clear day, warmth from the sun that hits snow is reflected back into the air and lost. And he believed that as glaciers grew, wind patterns would change and that this could lead to changes in oceanic currents.

Croll had many of the facts correct, but his timing was off. As geologists learned more about the coming and going of cold, they saw that the cycles from Croll's work did not match what they saw on the ground. Croll's math was out of kilter with the earth's behavior, and his ideas were discarded. But, as sometimes happens in science, his ideas were later resurrected. Milutin Milankovitch, a Serbian mathematician and engineer known as an authority on the properties of concrete, decided that his talents could best be used in developing a mathematical theory of the earth's climate. Working through the chaos of the First World War, spending part of that time as a prisoner of war, using new information on just how much energy the sun delivers to the earth, Milankovitch reworked what Croll had started. In 1920, he wrote *A Mathematical Theory of the Thermal Phenomena Produced by Solar Radiation,* a book with a title that doomed it to limited circulation. But among the few readers of the book was Wladimir Köppen. Köppen's daughter was married to Alfred Wegener, the man responsible for the theory of continental drift, who would later, in a historical footnote, die on the Greenland ice. Köppen and Wegener realized that Milankovitch's work could be extended to the distant past. They wanted him to run the calculations to six hundred thousand years before the present. The results were consistent with the history of the Ice Age as it was then understood. In 1941, Milankovitch published another book with another catchy title, *Canon of Insolation and the Ice Age Problem.* The Second World War broke out. The manuscript was at the printer when the Germans invaded Yugoslavia. The printing shop was flattened, but the manuscript survived more or less intact and was printed and distributed during the German occupation. From the memoirs of Milankovitch: "Our civilized existence had disintegrated into a life of hard grind."

Like Croll's work, Milankovitch's efforts were eventually dismissed. In a repeat of Croll's situation, new evidence suggested that the timing from Milankovitch's models did not match what had happened on the ground. But the parallel with Croll extended to the resurrection of Milankovitch's work. In 1976, a trio of scientists showed that the on-the-ground record matches a set of overlaid astronomical cycles: changes in the earth's orbit over a hundred-thousand-year cycle, changes in the earth's tilt over a forty-three-thousand-year cycle, and wobble of the earth's tilt over a twenty-thousand-year cycle combine to correspond to some degree with what is known about climate history during the hundreds of thousands of years of the Pleistocene Ice Age.

Circle the earth in late winter, and what do you see? In the

Northern Hemisphere, half the land is covered with snow, and a third of the ocean is frozen. We are in the midst of a warm spell, we are worried about global warming, but the fact remains that even in summer, whole regions remain covered with snow and ice. An area of land five times the size of Texas is in the permafrost zone, underlain by permanently frozen ground. If the mathematical predictions are right, we are at the tail end of an interglacial period, dramatically increasing its warmth with greenhouse gas emissions. But nevertheless we remain in what a geologist one hundred thousand years in the future would clearly recognize as part of the Pleistocene Ice Age. If the Ice Age does not die a natural death, and we do not kill it with greenhouse gases, renewed glaciation will come within a few thousand years.

This Ice Age at its worst, when what would become New York City was under ice, and woolly mammoths strolled over what would become great cities, was not as horribly cold as it might have been. For truly cold weather, one has to go back seven hundred million years, back to the time of Snowball Earth.

❄   ❄   ❄

It is September twenty-third. Back in Anchorage, leaves on the birch trees have turned yellow. It has been raining while I was away. It is raining now. Everything is soaking wet at the lower elevations. Everyone is speculating about a very snowy winter. "A few degrees colder," people say, "and this would be snow." In the mountains, at elevations above the road system, snow already covers the ground. A certain class of women, not quite able to wait for the really cold weather, already wears fashionable full-length coats. People are putting on their spiked snow tires. The spikes click and clack as they contact wet pavement.

My pet caterpillars, Fram and Bedford, stayed in the refrigerator while I was in Britain. They seem a bit subdued but otherwise fine.

I give them fresh willow leaves to eat. They crawl on the leaves, but they have no appetite. For them, it is close to winter. It is nearly time to freeze up.

Friends come to dinner. One of them brings me a copy of the *Anchorage Daily News,* dated September ninth. I had been driving on the ninth, crossing Scotland's glacier-formed landscape. "Lodge Owner Trapped by Lawnmower Dies," the headline says. Andrew Piekarski had been cutting the grass at his lodge thirty miles from Anchorage. He was on a riding mower. Nighttime temperatures in his area were dropping into the thirties. On a small hill, the mower toppled and pinned him to the ground. He lay under the mower all night, unhurt but trapped and slowly freezing to death. A state trooper was interviewed. "He couldn't get it off his legs," the trooper reported, "and he couldn't get out from under it and he died from exposure, from hypothermia." As Piekarski slowly froze, there would have been time for existential thinking. Before the hypothermia slowed his thinking, before the stupor, before the hallucinations of warmth, he must have considered the absurdity of his situation, the odd reality of deadly hypothermia before the end of summer.

I show off Fram and Bedford to my dinner guests. The two caterpillars remain listless. It is clear that they are ready to sleep. I have two jewelry boxes, and I line each with willow leaves. I put Fram in one box and Bedford in the other. After a brief ceremony, I put them in the freezer, behind the frozen peas and under a slab of salmon, condemning them to a snowball cocoon for winter.

❄   ❄   ❄

Reconstruction of past climates is not a simple business. Even today, three centuries after Daniel Fahrenheit developed the mercury thermometer, understanding recent global climate change leads to controversy. Which thermometers are reliable? Do thermometers near cities reflect local warming or real patterns in global temperature change? Do the number of measurements taken in the Northern Hemisphere outweigh and overwhelm those taken in the Southern Hemisphere? Is the average temperature the important number? Or the hottest day? Or the coldest day?

There are dozens of ways to reconstruct past climates. In western Europe, the timing of the grape harvest has been recorded for hundreds of years. Late harvests reflect cold, wet summers. Emmanuel Le Roy Ladurie, author of *Times of Feast, Times of Famine: A History of Climate Since the Year 1000*, wrote, "Bacchus is an ample provider of climatic information. We owe him a libation." Where records exist, grape harvest data can be backed up by grain harvest data. Another climate historian visited forty-one art museums to look at more than six thousand paintings dating from the beginning of the Little Ice Age. In the paintings, he saw a slow increase in cloudiness from the late 1400s until about 1750. Low clouds settled in after 1550 but cleared away around 1850.

Much can be said without written history or paintings. Tree rings can be measured, often stretching back a century or more. Corals, too, leave growth rings. In both cases, wide rings indicate good growing conditions, while narrow rings indicate cold. The chemical composition of ice cores says something about the climates when the ice formed. Cores pulled from Antarctic ice can span four hundred thousand years. Past sea levels say something. When it is cold, water is locked in ice, and sea levels drop. River channels extending out onto today's continental shelves speak of lower sea levels during past cold snaps and glacial periods. Beach terraces on hillsides speak of higher sea levels and past warm spells and interglacials.

And there is geology. There are erratics and moraines. There is bedrock scored by the action of ice. There are fossils of animals and plants with known temperature tolerances. There are sand wedges, originally formed as long-ago ice wedges, identical to those of today's Arctic. There are dropstones—stones carried by ice over a sea or a lake and dropped to the bottom when the ice melts, there deforming soft mud, leaving an unmistakable sign of ice and cold. There is topographical graffiti left by massive flooding that followed warm spells and broken ice dams. The Channeled Scablands of eastern Washington State were scrawled across the landscape when the ice dam that formed Lake Missoula melted thousands of years ago. Water from the lake ripped across what is now the northwestern United States. For a few hours, water flowed at a rate something like sixty-five times that at which today's Amazon River flows. During these few hours, soil disappeared and boulders were suspended like particles of clay. Gouges hundreds of feet thick were cut into the earth.

There is this inescapable fact: the farther back in time one goes, the more speculative climate reconstruction becomes. It is easy enough to find signs of the most recent glaciations of the late Pleistocene Ice Age, to see hints of its many glacial and interglacial swings, but dip farther into prehistory, and records are obliterated. Rocks are worked and reworked, clues are muffled, and fossils are scarce or even absent. The earth has been around for something like four and a half billion years. It took the first billion years for life to invent itself. This was simple life, more like a living slime than something biblical. It took another three billion years for more complex life to form, the kind of life that leaves abundant fossils, the kind of life with hard shells and bones and exoskeletons. Throughout this time, continents drifted. Tectonic plates rode about like loose barges, occasionally colliding, one forced downward and one upward. And there was erosion from wind and water and ice. The rocks enshrining the planet's history were recycled, cast away like dusty books remaindered from a publisher's warehouse.

Exactly how, then, does one reconstruct the climate of seven hundred million years ago or a billion years ago or two billion years ago? From work emerging over the past fifty years or so, but really coalescing over the past ten years and largely through the efforts and personality of one man, the idea of Snowball Earth has taken hold. It is science at its ugliest, when evidence is scarce and inconsistent, when speculation and ego and charisma mix with observations, when data are insufficient to unambiguously confirm or refute ideas. And for now, the speculation and ego and charisma of Paul Hoffman— backed up by a certain amount of hard-won data from remote locations in the Canadian Arctic, the Namib Desert, and Australia's Flinders Ranges—has won the day. The earth of seven hundred million years ago, Hoffman believes, was frozen from pole to pole, one more or less continuous, frigid, blood-thickening, numbing ball of ice: Snowball Earth.

His evidence is in the geology of Precambrian rocks. It is the same sort of evidence left by the Pleistocene Ice Age, but the rocks are much older and occur at all latitudes. The tropics, seven hundred million years ago, were not so tropical. The oceans were frozen over. One could have ice-skated between tropical islands. Across much of the earth's surface, a mercury thermometer would be of little use because the mercury would be frozen solid. Mean global temperature could have been something like minus sixty degrees. At the equator, temperatures would have hovered around negative ten, what Apsley Cherry-Garrard would have called forty-two degrees of frost. The planet would have been only slightly more hospitable than, say, Mars.

As early as the 1870s, people were finding scattered evidence of long-past ice ages, ice ages much older than Agassiz's original Ice Age. There was, for example, a paper by H. Reusch published in 1891 with the title "Skuringmærker og Morængrus Eftervist i Finnmarken fra en Periode meget Aeldre end 'Istiden' "—in English, "Glacial Striae and Boulder-Clay in Norwegian Lapponie from a Period Much Older Than the Last Ice Age." In 1948, the Antarctic explorer Sir Douglas Mawson spotted signs of ancient ice ages in Australia. "Verily," he told the Royal Geological Society of Australia, "glaciations of Precambrian time were probably the most severe of all in earth history; in fact, the world must have experienced its greatest ice age."

Others slowly built on the idea of an ancient ice age, but it took the ambition, ego, and entrepreneurial maneuverings of Paul Hoffman to bring the idea home. He has been quoted as saying, "Everyone's entitled to my opinion." During a scientific debate, he once challenged a Snowball Earth naysayer to "try me out in the Boston Marathon some year." And his own self-assessment, according to a biographer: "Gosh, I'm awful. I don't know how I'd react to me." But the history of science shows that new ideas take hold because they are pushed by strong personalities. In 1998, Hoffman, with a couple of coauthors, published "A Neoproterozoic Snowball Earth" in the prestigious academic journal *Science*. The idea and the people behind it were immediately attacked, both verbally and in print. The "Snowball" jargon had charisma, but it was a charisma that worked both ways. Opponents wrote papers with titles such as "The Snowball Earth Trip: A Neoproterozoic Snow Job?" and "Has Snowball Earth a Snowball's Chance?" Some talked of a "Slushball Earth," not quite as cold as Hoffman's Snowball and with oceans that were not entirely frozen over. Others rejected the idea entirely. Still others have gone in the other direction, delving farther back in time, suggesting that an even colder Snowball Earth circled the sun some two billion years earlier than Hoffman's did.

Hoffman has been awarded the Logan Medal by the Geological Association of Canada, the Miller Medal for outstanding research in earth science by the Royal Society of Canada, and the Alfred Wegener Medal by the European Union of Geosciences. He is a member of the National Academy of Sciences. He is a professor at Harvard. But he also has been accused of founding the Church of the Latter-day Snowballers. He stirs up bitter debate. It has been said that he sometimes alienates people who were once friends. He inspires passions, both positive and negative. He is, in short, a successful scientist. He has, if nothing else, made the world of science consider the possibility of a chilled earth, an icicle planet, one big ball-shaped skating rink in bad need of a tropical vacation.

❄ ❄ ❄

It is September thirtieth. Last night, it dipped to thirty-six degrees in Anchorage, but by noon it is sunny and nearly fifty. There are eleven hours, thirty-two minutes, and one second of daylight today, sunrise to sunset. That is five minutes and forty-one seconds less than yesterday, and five minutes and forty-one seconds more than tomorrow. By December twenty-first, the shortest day of the year, we will be down to about five hours of daylight. The sun, when up, will drift low in the eastern sky, nowhere near overhead at high noon, but rather just above the mountains, angling in, its rays scattered and pasting long shadows on what will by then be thick snow. My son and I drive out to Eagle River with the top down. He complains about the cold. Soon the convertible will have to be parked for the winter.

It has been snowing in the mountains again. Above us, what had been termination dust is now a shroud of snow. On the trail, well below the snow, we pass a hunter lugging out a Dall sheep. Under his load, the hunter is hunched over but moving fast. He looks as though he has been sleeping on the ground for several days. Although he looks bad, his sheep, draped over a pack frame and strapped in place, looks far worse.

Dall sheep, smaller relatives of the bighorn sheep that live in the Rocky Mountains, roam the mountain slopes winter and summer. It would be fair to say that sheep seem uncomfortable on level ground. It would only be a slight exaggeration to claim that a Dall sheep can stand upright on a vertical cliff. The rams, with their long curling horns, live in bachelor bands. The ewes, with stubby billy-goat horns, give birth in May or June, and by the onset of winter the young have weaned. Both rams and ewes butt heads now and again to maintain social order. They live in what is sometimes called "escape terrain," steep rugged slopes where they can elude predators. In summer, they eat the green fuzzy stuff growing between the rocks. In winter, they eat frozen grasses and lichens and moss where the wind has whipped away the snow on exposed ridges. They lose weight.

Just north of here, students at Wasilla High School are raising funds to help pay the hospital bills of their basketball coach, Jake Collins. In August, Jake and his fifty-three-year-old father had been sheep hunting in the Wrangell Mountains. They had driven a four-wheeler eighteen miles down a mud trail and then hiked in another six miles. Doing the steep terrain work of sheep hunting, they were at something like four thousand feet. The mountains were alive with ewes and lambs and, higher up, rams with full-curl horns. Jake went after one of the rams. His father watched from below. "There was some really gnarly stuff that had me nervous," his father later told a reporter, "but he got through the gnarly stuff." He watched his son shoot a ram. The ram dropped. Jake moved toward it but was thwarted by the terrain. He tried a different route. He was above the ram and tried to work down toward it but was stopped by a cliff. He went back up. "I think that third time was when he fell," his father said. When his father reached him, Jake was lying in a creek, bloody and unconscious, one eye swollen shut. Jake's father dragged him from the creek, took off his wet clothes, and redressed him in some of his own gear. Night fell. Freezing conditions are not unusual in the Wrangells even in the middle of summer. The wind blew. By morning, Jake's father decided he had to go for help. There were the six miles to the four-wheeler, then the eighteen miles of muddy four-wheeler trail, then twenty-five miles to help. It was night before a rescue helicopter was airborne. The pilot came in with night goggles. Thirty-five-mile-an-hour winds made the helicopter shiver. When they reached Jake, he was still breathing. It took some time to get him into the helicopter, and more time to get him to an Anchorage hospital. By then, his body temperature had dropped to eighty-eight degrees, the temperature at which shivering slows or stops, muscles stiffen, and the mind becomes cold stupid. Twenty days later, Jake came out of a coma.

My son and I stick to the valley floor. We watch salmon fanning the sand just above a beaver dam. We walk along a flat trail, below the treeline and the snow, and well below sheep country. We bend over a small, weed-choked pool of water and scoop some mud into a plastic bottle. The water feels close to freezing, like that of Prudhoe Bay in July. I put the bottle in my day pack. Later, I will put it in the freezer with Fram and Bedford, next to the frozen vegetables and a slab of salmon, and in spring, we will thaw it and see what comes out.

When I look up from here, certain slopes look almost skiable. It is late in the afternoon. Shadows have dusted the mountainsides, and a breeze carries the cold air downward, into the valley, onto this trail. Despite the day's sunshine, winter breathes down our necks.

OCTOBER

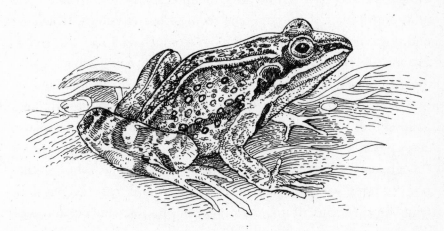

It is October third and forty-five degrees in Fairbanks. Patches of early-autumn snow lie scattered in the shadows of buildings and trees. On the University of Alaska campus, students wear jackets. There is a correlation here: when it is cold, people wear jackets. It seems momentarily plausible that the jackets cause the cold. People put on their jackets, and winter comes. They encourage winter, welcoming it as others would welcome sunshine.

Fairbanks is a cold-affected town. It was settled in 1901, when the skipper of the steamboat *Lavelle Young* decided that he could not go any farther upriver, toward the goldfields. He offloaded a man named E. T. Barnette near what would become the corner of Cushman Street and First Avenue. Barnette would become influential, but the place would have amounted to nothing had it not been for Felix Pedro. It was Pedro who found gold nearby just a few months after Barnette landed. For a short time, Fairbanks became the largest city in Alaska. And often it is the coldest. The average temperature in January is minus ten. The temperature has been known to go close to a week without breaking minus forty. Thermometers in Fairbanks know what it is to dip south of fifty below. And the people sometimes look weathered, too, pale and hardened. They are thicker than most people, more insulated for winter. With minus forty just around the corner, this is no place for anorexia. Beards are abundant and robust. Half of these people seem to live in cabins, which, by and large, does not mean a log cabin so much as a plywood shack. Having said that, when an Alaskan claims to live in a cabin, what is meant is never entirely clear. A cabin could have a dirt floor and log walls, or it could just as easily have five bedrooms, picture windows looking over a lake, and an alarm system tied to a remote response service that notifies the owner if the central heat fails. In Fairbanks, though, cabins tend to be on the shack side of things. Many do not have running water, in part because wells have to go deeper than empty pockets can afford, and in part because arsenic occurs naturally in the rock around Fairbanks.

The cold weathers more than just faces. Road surfaces are wavy from frost heave. Houses are slumped from thawing ground ice. Paint is dull and chipped, seemingly because of the cold but perhaps also because the summer is too short to be wasted on the business end of a paintbrush. But if the town and people are weathered, they do not seem to mind. "There's hardly any wind," they will tell you, looking on the bright side. For their health, they roll in the snow and jump into hot tubs. They make ice sculptures, including a life-size phone booth with a working pay phone. When cold fronts pass through, they pose in Bermuda shorts in front of thermometers. The frozen Chena River makes a perfectly good road.

Wandering around the university, I find a flyer advertising a bike for sale. The flyer is fringed with tear-off tabs offering a phone number. The bike comes with studded tires and thick Gore-Tex gauntlets on the handlebars, a far cry from the Draisine of Mary Shelley's time. The gauntlets protect the hands and forearms almost to the elbows. "Perfect year-round commuter bike," the flyer says. Almost anywhere else, this would be a joke. Here, all of the phone number tabs have been torn off. At forty below, it is easier to jump on a bike and go than to start and warm up an engine.

The university spreads out across a low hill overlooking Fairbanks. Stately buildings, most of them no older than the students themselves, hold lecture halls and laboratories. The International Arctic Research Center uses an entire building. It has a curved front and big dish antennas on its roof. Inside these walls, Russians and Japanese and Americans intermingle, talking about things such as frost deformation and shoreline erosion in a warming Arctic and construction of pipelines in frozen ground. The Japanese, though they have no Arctic of their own, helped pay for the center. The Russians are here because they know more about development in the Arctic than anyone else. First under the tsars and later under the Communists, they built gulags in the far north, and today they build pipelines and roads and operate a port above the Arctic Circle. In light of today's warming climate and melting sea ice, they see the Arctic Ocean as an increasingly accessible frontier, a resource basin, and with their experience, they may be the first to cross the starting line in the inevitable race for fish and oil.

I am here at the invitation of a friend, a professor at the university. A graduate student shows me a time-lapse video clip of soil freezing in a test chamber with transparent walls. Horizontal bands of segregation ice appear. The water in the soil moves toward the bands, and the soil between the bands visibly dries. Vertical cracks form. The column of soil grows taller. With the right music and juxtaposed against the right scenes of cars bogged down in snow, men with icy beards, and musk oxen puffing hot breath against a frozen northern landscape, I feel certain that this could be a popular short film, something for the Sundance Film Festival. The student sees no humor in this prospect. She plays the video twice, pointing out the features of a freezing soil profile. A man interested in the engineering challenges of a gas pipeline explains how frost heave can bend a buried pipe, and another man shows me how the pipe will be squeezed and deformed as the ground freezes and thaws around it.

I have dinner with friends in a restaurant near Cushman and First, where E. T. Barnette was dumped with his goods in 1901. No one mentions Barnette. We are, for the most part, busy eating. We are busy fattening up for winter. Although it is not yet truly cold, we huddle together like miners around a woodstove or muskrats in a snow-covered lodge. If I lived in Fairbanks, I would put on fifty pounds and sleep until June.

❄   ❄   ❄

Humans, as it turns out, cannot hibernate. Hibernation would kill us. NASA scientists, wanting to understand the effects of weightlessness, recently offered thousands of dollars to volunteers willing to lie in bed for up to several weeks at a stretch. The ideal job, perhaps, except for this: bone and muscle mass decreases, digestion slows, tissues become resistant to the effects of the body's own insulin, and control of blood sugar levels deteriorates. And this: humans, despite what they may want to do, cannot sleep for days on end. They need a drink of water. They must urinate. They require a snack. They want to get up and walk around.

Hibernation, once thought simple, is complex. It is sometimes argued that bears do not hibernate at all. They merely sleep, the argument goes, without entering a deep stupor and without a dramatic drop in body temperature. Some call this "winter dormancy" instead of hibernation, to distinguish between a bear and, say, a ground squirrel. But the difference between hibernation and winter dormancy is not clear. One blends into the other. And different bears fall into different levels of hibernation. The heart rate for active bears hovers around a hundred beats per minute, but for hibernating bears it may drop to forty beats per minute, or in some cases as low as eight beats per minute. An expert working on black bears in the Carolinas claims that his bears always watch him as he approaches their winter dens. Another expert working on black bears in Minnesota talks of falling into a den. His fall scared a cub. It was crying, immediately next to its mother, but the mother took more than eight minutes to awaken.

Hibernation, in its broadest sense, is winter inactivity. It is a way of bypassing periods of food scarcity, of skipping those times when the calories needed to stay warm exceed the calories that can be reliably gathered. Black bears, preparing for hibernation in the late fall, can gain thirty pounds in a week while for the most part eating nothing but berries and foliage and maybe insects. Brown bears—grizzlies—can pick up eight inches of fat, becoming, according to official information published by the Alaska Department of Fish and Game, "waddling fat" just before hibernation. Then winter hits. Food availability diminishes, snow slows down movements of foraging animals, and cold burns calories. Animals preparing for hibernation stop eating. They repair to a den—maybe a hollow tree, a brush pile, a crack in the rocks, a corner in a cave, a pit dug into the side of a hill. A black bear was once found denning in a tree hollow ninety-six feet aboveground. Grizzlies have been known to tunnel nearly thirty feet into the earth. For polar bears, only pregnant females hibernate. In Alaska, the polar bear den is a snow cave, dug on land or on the sea ice, sealed over and invisible after the next real snow.

The bear curls into a ball, its head resting between its forepaws, its back to the cold. Heart rate drops. Breathing slows. The digestive tract shuts down for the winter. Blood concentrates in the head and upper body. Body temperature drops nine degrees. The bear lives off its summer fat. Cholesterol levels skyrocket, but without causing heart problems. The bear will not urinate for months. Urea, normally jettisoned in urine, is reabsorbed through the bladder wall and processed back to amino acids and proteins. Likewise, calcium leaking from bones into the blood is recycled.

Certain Native Americans thought of den emergence as a form of rebirth. Males usually come out first. Females with cubs follow. Emergent bears might appear stiff, hobbling about like old men. They stretch. They sniff about. They yawn and scratch. They are thirsty but not immediately interested in food. When their appetites return, they dine first on roots and herbs, restarting a dormant digestive process. The mighty polar bear, often thought of as pure carnivore, has been seen pawing through the snow to feed on frozen salad just outside its den. Bears at emergence, at rebirth, may weigh as little as half their autumn weight.

It is not just bears that hibernate. It is ground squirrels and chipmunks and groundhogs and raccoons and skunks and prairie dogs. Some, such as raccoons and skunks, hibernate softly, like bears, in a deep stupor but without the dramatic body temperature drops of ground squirrels and groundhogs. Some hibernate alone, others in groups. Groundhogs—variously known as woodchucks, marmots, and whistling pigs—sleep in family groups. All wake occasionally to warm themselves and to drift into a softer sleep, a nonhibernating sleep that allows them to dream.

Birds, for the most part, migrate. Until the 1940s, scientists believed hibernation to be as unbirdlike as carrying an umbrella. It was then that biologist Edmund Jaeger, wandering in the Chuckwalla Mountains of the Colorado Desert, watched poorwills. Poorwills are nocturnal birds, seven inches long, that feed on flying insects. "Where do they go in the winter?" Jaeger asked a Navajo boy. "Up in the rocks," the boy replied. The Hopi, too, knew the poorwill, and in their language called it "the sleeping one." In winter, Jaeger found what at first seemed to be dead birds in the rocks, but when he picked them up, they flew away. He watched one that stayed in torpor for eighty-five days. He pushed a thermometer up the cloaca of a torpid poorwill and found its body temperature close to that of the air. Its pupils did not react to light. He found no heartbeat. The bird did not seem to breathe. But in spring it flew away. Its awakening, its rebirth, coincided almost perfectly with the reappearance of flying insects. In a 1949 paper, Jaeger wrote, "I take it as evidence that the bird was in an exceedingly low state of metabolism, akin, if not actually identical with hibernation, as seen in mammals."

Later, biologist Jon Steen watched titmice and finches. With enough food, the birds shivered through the night, their feathers puffed up, their heads tucked under their wings, their aura, to the extent that birds project an aura, pathetic. But hungry birds entered a nightly torpor. Their body temperature dropped. They have since been called "daily hibernators." Biologist and author Bernd Heinrich has written, "Physiologically there is no distinction between hibernation and daily torpor."

Reptiles and amphibians are not exempt. Snapping turtles — air-breathing reptiles — can lie under the mud beneath the frozen surfaces of lakes for more than four months at a stretch. The Manitoba toad of Minnesota summers near flooded depressions left behind by Pleistocene glaciers, but in winter it hops upslope and burrows into gopher mounds. Two University of Minnesota biologists tagged hundreds of toads with radioactive chips and followed them through the winter. The toads, underground and without food, in amphibian semistupor, gradually burrow deeper through the winter, staying ahead of the ever-deepening frost line. Somewhere around four feet down, they stop digging. By spring, their body temperature is just above freezing, not much warmer than that of a hibernating ground squirrel. But unlike the ground squirrel, the toad will not shiver itself back to warmth. It will have to wait for its tunnel to warm, and then it will crawl out of its burrow, looking for sun.

The wood frog is to the Manitoba toad as Maine is to Florida. The wood frog's habit of overwintering frozen, with ice in its veins and between its cells, was not understood until 1982. Naturalist John Burroughs, walking in the woods of New York in December 1884 — nearly a hundred years earlier and nearly seven decades after the Year Without Summer — heard a frog calling from beneath the leaves. He lifted the leaves. "This, then," he wrote, "was its hibernaculum—here it was prepared to pass the winter, with only a coverlid of wet matted leaves between it and zero weather." Burroughs at first believed this to be a predictor of an easy winter. "Forthwith," he wrote, "I set up as a prophet of warm weather, and among other things predicted a failure of the ice crop on the river.... Surely, I thought, this frog knows what it is about, here is the wisdom of nature." The frog, he believed, would burrow into the ground if a cold winter was ahead. But he was wrong. Two feet of ice formed on the nearby Hudson River, and it was still bitterly cold when Burroughs went back to look for his frog in March. The leaves of the hibernaculum were frozen. He peeled them back, and beneath he found frozen ground. Between the frozen leaves and the frozen ground lay his frog. "This incident convinced me of two things," he wrote, "namely, that frogs know no more about the coming weather than we do, and that they do not retreat as deep into the ground to pass the winter as has been supposed."

When handled, Burroughs reported, the frog blinked. In this he was mistaken, for frozen frogs do not blink. The idea that a frog could spend the winter frozen was so outlandish that Burroughs appears not to have considered it. He saw what he believed because he could not believe what he saw.

Flash forward one hundred years to when a physiologist named William Schmid found a wood frog under the winter leaves in Minnesota. The frog was frozen. Schmid thawed it out and watched it come to life. In 1982, he published "Survival of Frogs in Low Temperature." So far, four frog species are known to overwinter in a frozen state. To be clear, these are not frogs that are cold, but frogs that are literally frozen. Pick them up, and they are as hard as ice. They are, in fact, largely ice. Almost two-thirds of their body water may be frozen. Ice crystals form between their cells and throughout their body cavities, but the cells themselves, protected by high concentrations of glucose, do not freeze. In this state, the frogs can survive body temperatures as low as eighteen degrees.

Bugs are stranger than frogs. Tent caterpillar eggs are full of glycerol, a form of alcohol that acts as an antifreeze. Caterpillars—my Fram and Bedford—simply freeze solid. Take an African desert fly, dry it out, throw it in liquid helium at temperatures below minus 450 degrees, warm it up, and pour some water on it, and it will demonstrate what it is to be a survivor.

And there is the trick of supercooling. Water, it turns out, has a sharp and consistent melting point. Warmer than thirty-two degrees, ice becomes liquid water. But chill water below thirty-two degrees, and it may still be liquid water. In this supercooled state, the liquid is unstable. Add a speck of dust or a snowflake—nucleation sites for ice formation—and ice crystals will grow. That, though, understates the process. Supercooled water, once it goes, goes quickly. It flash freezes. The ice crystals blossom. They explode into being. The trick to survival through supercooling is to avoid anything that might trigger flash freezing. To flash freeze is to die. Yellow jacket wasp queens latch onto the underside of a leaf or a piece of bark and then hang suspended, their body temperature dropping as low as four degrees. Supercooled, they do not freeze. But tap them, or let a snowflake hit them from above, or a drop of water, and they turn to ice.

Bear biologist Lynn Rogers was walking one day in January. Quite possibly, he had passed the burrows of toads earlier that day. He likely had trod past frozen wood frogs. He certainly would have been in the company of insects with antifreeze in their tissues, of frozen insects, and of supercooled insects. Then he crawled into a bear's den. "On January 8, 1972," he wrote,

> I tried to hear the heartbeat of a soundly sleeping five-year-old female by pressing my ear against her chest. I could hear nothing. Either the heart was beating so weakly that I could not

hear it, or it was beating so slowly I didn't recognize it. After about two minutes, though, I suddenly heard a strong, rapid heartbeat. The bear was waking up. Within a few seconds she lifted her head as I tried to squeeze backward through the den entrance. Outside, I could still hear the heartbeat, which I timed (after checking to make sure it wasn't my own) at approximately 175 beats per minute.

❋   ❋   ❋

It is October thirteenth and thirty-six degrees in central Pennsylvania. A Friday the thirteenth cold front, invading from the northwest, jumping the Canadian border without hesitation, decimated yesterday's T-shirt weather. The front dumped two feet of snow near Buffalo, New York, 150 miles from here. The *Washington Post* ran a photograph of a Labrador retriever bulldozing through the snow, its legs invisible beneath the white stuff, its nose iced over, the look in its eyes one of joyful confusion. The *New York Times* ran a front-page color photograph of cars buried in snow and people in winter coats, shoulders hunched, walking away from the camera, subliminally headed south. In Minnesota, the same weather system dropped temperatures to twelve degrees. It is not supposed to be twelve degrees in October, not even in Minnesota. "Don't forget the School Children's Blizzard," the weather is saying. "Remember Greely. Remember who is calling the shots."

Officially, I am here for a seminar on underwater sound, the sort of educational opportunity that only a large university can offer. Unofficially, I am far more interested in learning through experience and observation. I want to learn more about cold and to see firsthand how temperate zone students respond to it. I drive across the Pennsylvania State University campus. The students are slimmer than those in Fairbanks, and clean-shaven: more primped, less insulated.

Their warm clothes are formfitting, lacking the space underneath for three sweaters. In general, the students here are less prepared for an ice age, but at least one of them, graduated now and moved on, was a cryophile, a lover of cold. While here, he learned of a place just outside town known variously as the Barren Valley, the Scotia Barrens, and just the Barrens. With his roommate, he nailed a thermometer to a pine tree in the heart of the Barrens. The two of them drove the four miles out from campus once a month. "It is amazingly quiet, amazingly clear, and amazingly cold," he once wrote. When it was thirty degrees on campus, it was eight degrees in the Barrens.

Below-zero dips in temperature occur thirty times more often in the Barrens than on campus, and below-freezing temperatures occur twice as often. In the daytime, the Barrens warm up. By midafternoon, the Barrens are as warm as the campus. But at night, temperatures plummet.

The area, a narrow valley between steep hills, was called the Barrens because the soil was too sandy to support productive farming, a common naming practice. It was mined for iron ore late in the nineteenth century. The forest that had stood there was cut down and the wood turned into charcoal. The sandy soil was exposed. Scrubby plants grew into sparse patches of trees. Today houses encroach on the Barrens, but the heart of the place is owned by the state and managed for game. Heat escapes from the dry, sandy soil as soon as the sun dips below the hills. Cold air tumbles downhill into the valley and settles in for the night. A temperature of forty below was once measured in the Barrens—forty below zero in central Pennsylvania.

I leave campus at seven thirty in the morning. It is thirty-six degrees. Ten minutes later, I am in the Barrens. A possum rests peacefully in the road, its gray fur dusted with frost. I open my windows, enjoying the cold. At a sign that says GROUSE HUNTING AREA, it is thirty-four degrees. At a shooting range a few minutes down the road, it is thirty degrees. The forest is still recovering from the work of woodcutters and charcoal production. The trees are uniformly young and scattered. My left ear, exposed to the full blast of wind coming through the window, is now comfortingly numb. In the heart of the valley, still fewer than ten miles from campus, it is twenty-six degrees, a full ten-degree drop from campus. My hands are stiff on the steering wheel. My left ear is beginning to hurt. I turn the car around, my curiosity satisfied, ready to migrate back to the warmth of campus.

❄   ❄   ❄

The poorwill—to the Hopi people, "the sleeping one"—hibernates through the winter, but it is an exception among North American birds. Another hundred or so of the continent's birds tough it out through the winter, feeding frantically to supply the calories needed to stay warm. The other 550 species that breed north of Mexico migrate.

A migrating bird might fly and fly and fly some more, past cities, above browning cornfields and forests, over vacant Gulf of Mexico waters, a night and a day and a night passing in the air, a few ounces of feathers and flesh and aerodynamics, finally landing, wasted, almost nothing left. Or if it is, say, a Clark's nutcracker—a gray jay with a white face—it might go no more than a few miles, like a moose moving from a mountainside to a neighboring valley. If it is a mallard, it might hop south, moving from pond to pond one step ahead of the freeze line. Many birds stop in the southern United States. More than two hundred species go on to the beaches and coastal forests of Mexico. Others head for the Caribbean. A few dozen make it as far as the Amazon, overwintering with parrot friends. A few—the barn swallow, the upland sandpiper of the American grasslands, the Swainson's hawk—make it to the Pampas of Argentina. Some change their migratory habits from year to year, wintering wherever they find the right balance between availability of food and the lessening of calorie-sapping cold. Great gray owls, very much northern birds, head south only if they get hungry enough. They showed up south of their normal range in the winter of 1978 and again in 1983. One made it as far south as Long Island. There, well outside their normal range, they were sometimes seen feeding in broad daylight, not behaving as good owls do. They were hungry.

Aristotle believed that swallows hid underwater during the winter. He also believed that worms came from horsehair, and he thought that the European redstart transmogrified into the old world robin. For centuries, others saw the world through glasses tinted by Aristotle's errors. As late as 1555, Olaus Magnus, archbishop of Uppsala, wrote of swallows, "They cling beak to beak, wing to wing, foot to foot, having bound themselves together in the first days of autumn in order to hide among the canes and reeds." But Aristotle knew more than his widely publicized mistakes might suggest. "Others migrate," he wrote, "as in the case of the crane; for these birds migrate from the steppes of Scythia to the marshlands south of Egypt where the Nile has its source....Pelicans also migrate, and fly from the Strymon to the Ister, and breed on the banks of this river." Other ancients, too, knew of migration. Homer said that cranes "flee from the coming of winter and sudden rain and fly with clamor toward the streams of the ocean." The Old Testament reports "the stork in the heaven knoweth her appointed times; and the turtle [dove] and the crane and the swallow observe the time of their coming."

We continue to learn about migration. For example, each year the spectacled eider, one of the nation's most beautiful ducks, disappears from the summering grounds of the Arctic and reappears each spring. Where does it go in the winter? Prior to 1994, no one was sure. For all anyone knew, it could have overwintered underwater, bill to bill and wing to wing. In 1994, biologists followed the signal of a transmitter implanted beneath the skin of a spectacled eider. They found the eider, and more than a hundred thousand of its cousins, secure in the pack ice, jammed together in a pond of open water surrounded by thick ice and bathed in Arctic winter darkness, with air temperatures of thirty below. The eiders' collective motion and body temperature, it seemed, contributed to the maintenance of the open hole that was their home. They overwintered by feeding and paddling about in that.

There is more to be learned. There are, for example, physiological adaptations. Not unexpectedly, birds put on fat, but in some cases nonessential organs shrink. Just before migration, the bartailed godwit becomes fifty-five percent fat, but its kidneys, liver, and intestines shrink. Then it flies nonstop at something like 45 miles per hour for days on end. The speed and exact route of many birds are not known. Migrating sea ducks tracked by radar in the Arctic fly at more than 50 miles per hour. A dunlin—a long-beaked shorebird—was once clocked at 110 miles per hour, passing a small plane. Other unknowns: How do they cope with man-made obstructions? How do they respond to the lights and noises of cities and ships and smokestacks? Do flashing lights warn them away or just confuse them? In 1998, migrating Lapland longspurs came upon an antenna tower in the Kansas fog. Apparently confused by the tower's blinking lights, they circled it, again and again and again. They ran into guy wires, into the tower itself, into one another. Before it was over, ten thousand were dead on the ground.

Bird banding has been to avian biologists what the telescope has been to astronomers. A band—a metal or plastic tag—is clipped to a bird's leg, pinned to a wing, or placed as a collar around a neck. Each band has a number and an address. The bird might be found dead, or shot, or captured by other banders, giving up information on its movements. Henry IV of England banded falcons at the beginning of the Little Ice Age. Duke Ferdinand banded a heron in 1669. John James Audubon is said to have been the first bander in the United States. Today a single banding station, staffed for the most part by volunteers, might band a thousand birds in a week. The U.S. government issues one and a half million bands each year. Something like sixty-three million birds have been banded, and just under four million bands have been recovered. A letter from China accompanying the return of a pintail duck band reads, "I feel very glad to wrote you. I did not know you and you did not know me. Who introduced I to you? It is your pigeon. She Flew To China. What a far way she flew! It is marvelous." Less helpfully, other bands have been returned with notes asking for recipes.

A returned band will say something about where a bird went but not how it got there. Bird navigation is no simple matter. Navigation skills are both learned and instinctive. Birds follow rivers and shorelines. They use the sun and the stars. They hear breaking surf. They may detect differences in air pressure. Some have magnetite in their nasal cavities—built-in magnetic compasses. Experimenters have done odd things. One looked at bird behavior in a planetarium, confusing his birds by turning off stars and star groups and entire portions of the sky. Another put electromagnets in birdcages. After all of this, nothing is completely clear. The truth behind migratory navigation defies generalizations.

A flock of migrating birds is more than the sum of the individual birds, and less. Free will seems lost. Some combination of instinct and memory and groupthink drives the beat of wings away from the cold and toward reliable food. Birds flew some of these routes before men hunted with stones. Many knew the sting of the nineteenth-century market hunters, feeling the flak sent up from shotguns mounted on the bows of punts like batteries of antiaircraft guns strung along the nation's flyways. Birds have seen the lights of cities flicker on. They have run into buildings that grew out of prairies. Resting in trees or on the ground, birds have learned of the dangers of house cats: one hundred million birds feed themselves to tabbies every year.

Most recently, birds have discovered microwave communication

towers, running into them at full migratory speed, no doubt thinking, *This wasn't here last year,* just before abruptly stopping beak-first,
skull-second, bones shattered, then falling to the ground, migration
finished, game over, lights out.

❄   ❄   ❄

It is October twentieth and fifty-four degrees in Anchorage. In my
absence, the snow has crept farther down the mountains, but for
now it is still above the streets and houses and office buildings of
the city. I am disappointed. It has been too long since I last felt snow
under my feet. I drive to Flattop, a little mountain just outside town
with what is reputedly the most hiked trail in all of Alaska. I am here
to walk in snow, but also to look for ground squirrels, thinking that
this might be my last chance to see them before they hibernate.

Within ten minutes of the trailhead, at something like twenty-
five hundred feet above sea level, I see hoarfrost, water vapor frozen
into intertwined white crystals on the surface of the ground. The
willows along the trail are nearly leafless, and the crowberries hang
from stems that have turned from rich green to dry reddish brown.
A bull moose forages on the slope beneath the trail. Soon the moose
will find its way into the valleys, where it will stand in the winter
cold, gnawing bark from shrubby trees. It will lose its antlers. When
the leaves are gone and there is nothing to eat but bark, moose
might spend six hours feeding and twelve hours ruminating. Eating more to stay warm is pointless, because they cannot digest their
food as quickly as they can feed. They lose weight. In a bad winter,
many will starve. But it must be a very bad winter. Their metabolic
rate does not increase until temperatures drop below minus twenty.
As often as not, their problem is staying cool. Winter-acclimatized
moose are known to start panting as temperatures rise toward the
freezing point.

Within twenty minutes of walking, there is snow on the trail—

white dust scattered on the surface of dark mud and caught in the leaves of crowberries. Here, at twenty-six hundred feet above sea level, it is ten degrees colder than at the trailhead. The clouds open for a moment, and the sun bathes Anchorage and Cook Inlet. I pull on a hat. A few minutes later, the clouds close up again, and big slow-moving snowflakes drift down, sticking to my shirt. At around twenty-eight hundred feet, I scan a slope of splintered bedrock and ice-shattered boulders, looking for ground squirrels. They are common here, among the rocks. In summer, they stand on their hind legs, imitating prairie dogs, watching tourists panting and struggling up the mountain. The ground squirrels will overwinter in burrows hidden beneath the rocks. It is in these burrows that they fall deep into hibernation. It is here, during winter, that their body temperature drops to just below the freezing point of water, close to thirty degrees. The ground squirrel's blood is more than ten degrees colder than that of a hibernating chipmunk. In the laboratory, a ground squirrel blood sample would freeze at thirty-one degrees, but within their frigid sleeping bodies, ice does not form. Their blood is supercooled—it is below the freezing point, yet not frozen. Supercooling occurs when there are no nucleation sites, no place for nascent ice crystals to get a toehold. Formation of a single ice crystal in the supercooled blood of a hibernating ground squirrel could flash freeze the little beast. The ice crystals would rip through cell membranes. The squirrel would die.

Here, too, in the burrows beneath these rocks, as often as every two weeks through the long cold of winter, the squirrels warm up. They are no different from other hibernators in that they need their sleep. They burn away fat reserves to warm up enough to sleep. An electroencephalogram of a cold squirrel shows a quiet brain, but when they warm up enough to sleep, the same electroencephalogram will show the patterns of dreaming. Dreaming of what? Of spring? Of succulent shoots and fresh flower buds? Of mating? For about a day, they warm up, and then they cool off again, quickly dropping to near freezing, and then somehow maintaining their body temperature at just above freezing. How do they do it? How does a nearly frozen brain, its neurons less nimble than cold molasses, tell the little rodent that it is time to wake up? No one knows.

In the cold air, I can hear Anchorage traffic. The combination of cold air and snow plays tricks with sound. If I turn my head slightly, or move a bit, the traffic noise is replaced by the sound of fast water coursing through the south fork of Campbell Creek, some eight hundred feet below me. Sound travels quicker through cold air than warm, but snow muffles sound. Between traffic and flowing water, I listen, too, for golden-crowned sparrows, for their tweeting song that sounds like "three-blind-mice, three-blind-mice," high-pitched and repetitive. But the little birds are gone, headed south toward Washington and Oregon and, for some, as far as California and the Baja peninsula. I check my thermometer. At this elevation, somewhere close to three thousand feet, it is thirty-four degrees. It will drop well below freezing tonight. There are no squirrels. They are already curled up in their burrows, bivouacked. It is snowing hard now. The city is lost behind a curtain of snow, and the mountains around me appear through a fog of slow-moving white flakes.

The last few hundred feet of the mountain are as steep as the steps of a lighthouse. The trail this high is covered with snow, and under the snow, from the trampling of other walkers, is ice. Eventually, I am clambering, grabbing rocks to avoid slipping. The snow — its coolness, but also its texture, crunchy but soft — feels good on my hands. Just below the summit, my right foot slips, and then my left, and it is only my hands that keep me from falling. That is how it is with snow-covered ice. Blink and what feels like firm footing becomes slicker than oil. My toes find a purchase, and I look down. I would not die if I fell from here, but I would be bruised and maybe broken. My fingers are numb. I have no need to crawl up the last twenty feet to the summit, so I turn and inch my way downward. Thin snow on top of ice has left the trail so slippery that I am trembling now and crab-walking downward. In places, I slide shamefully along on my behind. The snow eases for a moment, and the city materializes in the growing twilight. Through the lessening snow, Cook Inlet glows orange and blue with steeply angled rays of evening autumn sun.

NOVEMBER

It is November fifth in Palawan, an island in the Philippines, 500 miles north of the equator and 250 miles south of Manila. It has dropped below twenty degrees in Anchorage, and the North Slope has touched zero, but here in Palawan, it is eighty-two degrees. My migratory compass has, for the moment, acted birdlike, bringing me to warmer climes. Somewhere here on the island, the arctic warbler should be overwintering, just arrived from Alaska. If these birds are here, they are likely worn out, still recovering from a self-propelled migration, adjusting to the heat, coping with a new suite of predators, and recuperating from their own version of jet lag. But I see none of them. Instead, I see resident birds—a tropical kingfisher gliding low over blue water, a slender white-bellied woodpecker working the trunk of a coconut palm, a Philippine sea eagle riding thermals up the side of a limestone cliff. These birds will never know the feeling of snow between their toes. They will never know sea ice.

Underwater, I hold my breath and listen. The water temperature hovers around seventy-two degrees, and the sea is alive with the clicking of shrimp, sounding something like grains of sand and gravel awash in surf. In the distance, I hear the motor of a fisherman's banca, a thin canoe-like craft with twin outriggers for stability and a jury-rigged gasoline motor thumping through the waves. My companion says she can hear damselfish. I listen again. And there it is, a low-frequency grunt, an unmistakable burp just at the edge of my hearing. And then another answering the first. Now, my ears attuned to the right frequency, I hear a chorus of burps, one after another, burp after burp maybe saying something meaningful in damselfish-speak or maybe just the sound and the fury of three-inch reef fish.

These little fish are stuck here in the tropics. They die in colder water. A near relative, the blacksmith, which broke away from its tropical cousins some nine million years ago, lives as far north as Monterey Bay, California. The difference between the two? Enzymes. Enzymes are protein molecules, but unlike other proteins, enzymes are there to encourage reactions important to life. Bring two molecules together without an enzyme, and they may eventually react. Bring the same two molecules together on the surface of an enzyme, and a millisecond later they have reacted. To say that enzymes speed up reactions would be to downplay their importance. They rocket reactions into hyperdrive, increasing reaction rates several million times. And they are selective. They bring together certain molecules but ignore others. More than four thousand biochemical reactions are known to be mediated by enzymes. Enzymes are nothing less than the linchpins of metabolism.

But here is the catch: different enzymes work best at different temperatures. Take a damselfish enzyme that works best in the Philippines, make a few subtle amino acid changes across nine million years of evolution, and you have an enzyme that works best in northern California. *Molecular Biology and Evolution,* an academic journal, reports on the differences between damselfish and the related blacksmith: "Enzyme adaptation to temperature involves subtle amino acid changes at a few sites that affect the mobility of the portions of the enzyme that are involved in rate-determining catalytic conformational changes." Sketches of fish enzymes, twisted and retwisted ribbons spiraling across the page, show the difference between the enzymes of the tropical damselfish and the blacksmith. The difference as illustrated is not profound. The difference in the real world is the difference between life and death.

Anything affecting temperature affects the efficiency of enzymes. So do certain poisons. Temperature—the wrong temperature—acts like poison. It is not so much an issue of cold taking a single enzyme out of commission as one of cold disturbing the synchronous behavior of an orchestra of enzymes, leaving one playing too slowly, another too fast, and another barely playing at all, and in the end reducing the symphony of metabolism to the cacophony of malaise and death.

Temperature limits the ranges of species. Tropical damselfish are tropical not because they like warm water so much, but because they need warm water. By impairing enzyme performance, cold water kills tropical damselfish, or slows their growth or stops them from reproducing. The same is true for the corals out here, and the sponges, and the clicking shrimp. Move to the north or south, and the corals and sponges and shrimp are gradually replaced by other species with other enzymes that work at other temperatures. First one species drops out, then another, then another. They all play the odds, balancing enzyme efficiency against other risks. The enzymes of a fish may work best at eighty-one degrees, but if the enzymes of its predators work best at eighty-one degrees, it may be better off a bit farther north. It might do the same thing to avoid competitors. It may be better off taking its chances in places where the occasional cold snap could wipe it out. Or it may be better off warm-blooded, fighting off the cold outside to keep those enzymes at peak efficiency. It would need more food; it would need more insulation;

it would need adaptive behaviors and down jackets and baseboard heat and maybe an electric blanket. But it could live almost anywhere, while these damselfish are condemned to the tropics.

When I tire of listening to grunting damselfish, I stand in the water. Facing shore, I look at coconut palms growing just beyond white sand. Farther back, in the hills behind the beach, macaques play in jungle branches. I turn to watch the swells coming in from the South China Sea. They break on the reef face. A surge of water, the remains of a broken wave, advances across the shallows, then retreats, leaving white foam behind to slip slowly back. The sun hangs low on the horizon. At this latitude, sunset comes quickly. I can almost see the sun move as it sinks into the sea. This is the same sun that will rise in just a few hours over my home in Alaska, but there it will rise slowly, seeming to skim along the horizon, reluctant to show itself to snow-covered black spruce and frozen tundra.

❄ ❄ ❄

There was a time during the history of the science of ecology when a few obvious observations could become a rule. Today ecology relies on advanced statistics, on multivariate models and randomization techniques and the use of Greek letters in place of actual words. But a hundred years ago, it relied on narrative descriptions and observations. People like Karl Bergmann—working in the 1840s at the same time that Louis Agassiz was studying glaciers—noticed that animals in cold climates were bigger than their warm-climate cousins. Siberian tigers were bigger than Bengal tigers, and Bengal tigers were bigger than tigers in equatorial jungles. Northern badgers were bigger than southern badgers. White-tailed deer in Michigan were nearly twice the size of those in Nicaragua. Somewhere along the line, Bergmann's realization that cold-climate mammals were often bigger than their warm-climate cousins became known as Bergmann's Rule. Working with the chaos of nature, ecologists cling to patterns. They ignore or excuse exceptions. Most ecologists were not troubled by the fact that Arctic brown bears were far smaller than the brown bears of southern Alaska. An animal in a northern climate, the explanation goes, should be bigger, because a bigger animal presents less surface area for every ounce of weight. As the size of a cube or a sphere or an irregular shape — the shape of a bear or a moose or a caribou — increases, its surface area increases less than its volume. For every pound of body weight, for every ounce of muscle and fat and liver and lung and heart, a big animal has less skin exposed to the cold than a small animal. A big animal holds on to its heat more effectively than a small one, all else being equal.

Bergmann's Rule is not so much a rule as a pattern. And there are other patterns. Here you are in an oak forest, and a bit farther north you are in a spruce forest, and a bit farther you are in a treeless plain with frozen ground. These are the biomes of ecological maps of the world. The oak forest is temperate deciduous forest, the spruce forest is taiga, and the treeless plain is tundra. Biomes have clear boundaries on maps of the world, but the boundaries blur on the ground. The taiga fades to scrubby spruce trees, their enzymes failing them and their roots struggling to find a way through frozen soil. The spruce trees become more scattered. Seedlings might take hold for a year or two in a patch on the south-facing slope of a hill or along a river, then die back when a brutally cold, dry gale howls in from the north, blowing them over or sucking away the moisture that keeps them alive or sand-blasting them with crystals of wind-borne ice. Or the boundary might move way north, as it would have during the Medieval Warm Period, and then south, as it would have during the Little Ice Age. Stragglers, unable to reproduce in current conditions, their flowers not maturing or their pollinators no longer present, might hang on long after the climate has changed, little islands of out-of-place spruce or pine or oak trees.

The ranges of species go where the species work best, destined by the character of their enzymes, destined by how well their enzymes work at different temperatures. But also: Who will graze on my leaves? Who will eat me? Whom will I eat? Is there space for my nest? Is the soil right for my burrows or my roots? Who will drive me away? Puffins became scarce around Britain after 1920 not because of the air temperature, but because the fish they ate followed a shift in water temperature. The birds followed the fish. When water temperature shifted again around 1950, the fish returned, and with them the puffins. The lives within the biomes are interwoven, and if one species can go no farther because of the temperature, it may affect another species, and another, and another, until it appears as though there is some definite boundary and that everything responds in concert. But zoom in on the map, look a little closer, and the boundaries blur. Brown bears live in tundra and taiga and temperate deciduous forest. Caribou migrate across biome boundaries. The red fox, the tiger, the wolf, the wolverine, and the raven all cross biome boundaries as if they did not exist, as if they have never read an ecology textbook or studied a biome map.

During the Little Ice Age, the cod fishery in Iceland—a fishery that had been active for centuries—failed as certain fish moved south to warmer water. The Dutch prospered when cod appeared in the North Sea. The range of man, of *Homo sapiens,* changed, too. Eskimo hunters showed up in Scotland. Iceland's population fell by half, and the Norse abandoned settlements in Greenland, or their inhabitants simply died. And then the weather broke. The Little Ice Age gave way to a warming trend, perhaps helped along by carbon dioxide dumped into the air from the fires of man—first wood, then wood and coal, then wood and coal and oil, all burning at once, all contributing to the slow construction of a global greenhouse. Bering headed north, along with De Long and Greely and Peary and Cook. Amundsen and Scott and Shackleton headed south. Some of them retreated alive, while others perished like the Norse in Greenland, like spruce trees rooted too far north, too far into the polar zones, too far into the cold, and buffeted by deadly winds.

❄ ❄ ❄

It is November twelfth and eighteen degrees in Anchorage. My migratory timing has sent me back early, back to the cold, back to the coming of winter. Spring remains a long way off. The official start of winter is weeks away, but a thin veil of snow covers trees and houses and grass. The lakes around Anchorage have frozen, and where wind has blown the snow clear, ice crystals as big as flower buds blossom across their surfaces.

Here at Powerline Pass, above Anchorage, brown grasses poke out through thicker snow. Mixed with the grasses, fat stems of flower stalks, brown and dry, have split open just above the snow, their seams burst by expanding ice. Thigh-high shrubs — leafless, their bark scraped by moose teeth — stand with their heads and shoulders above the snow. Soon enough, the shrubs will disappear until spring, sleeping under the snow, away from the bitter cold and the wind that will careen down the surrounding slopes and through this mountain pass. The boughs of white spruce, sloped downward and out, already shed snow, and the snow piles up around them, while the ground immediately beneath them remains bare, protected by the umbrella of the boughs themselves and warmed by dark, sun-absorbing bark.

We are skiing, my son and I. For him it is the first ski of the season. He is awkward with his summer legs and new boots. He falls twice, but ten-year-olds bounce well, and his falls hinder neither pace nor exuberance. It is good, at times, to be in the cold. It stimulates the senses and the mind. Douglas Mawson, the Antarctic explorer, once wrote, "During the long hours of steady tramping across the trackless snow-fields, one's thoughts flow.... The mind is unruffled and composed and the passion of a great venture springing suddenly before the imagination is sobered by the calmness of pure reason." My son would express it differently. He would say it is fun. It might seem that his thin frame would chill easily, but he routinely sheds layers and complains when his teachers make him wear gloves at recess. Now he asks me to carry his hat and outer jacket, and then he rushes down the trail, skis flopping in rapid clumsy steps, poles gripped ham-fisted too far from his body, dark hair fluffed straight up by the discarded hat.

Stone Age pictures of skis and skiing appear on the walls of caves and on rocks. Four thousand years ago, the nomadic Sami, now incorrectly known as Laplanders, skied after the reindeer herds of Scandinavia. The word "ski" comes from the Norwegian *skith*, which means, literally, "wooden stick." Until very recently, skis were wooden sticks. Specifically, they were birch sticks. Where my son and I ski now, we are too high for birch. There is only willow and spruce and alder. Our skis are made of a composite material. At this elevation, composite trees are at least as scarce as birch.

The physics of skiing is surprisingly complex. The pressure of a ski on snow combined with the movement of the ski creates friction, and friction generates heat. It is occasionally said, often with great authority, that the pressure of a ski lowers the melting temperature of the underlying snow and that movement of the ski adds heat through friction, creating a microscopic layer of liquid water that lets the ski slide. This explains why skis drag when temperatures drop. At minus thirty, when the snow is too cold to melt into microlayers of liquid water, the skier ceases gliding and begins to scrape. But it turns out that the idea of a microlayer of water between snow and ski may not be right. Something else may be going on. It may be that the molecules of frozen water at the surface of the snow, while still frozen, are not bound as tightly to the crystal lattice beneath. The surface molecules cannot grip other molecules above their heads and under their feet, so they slide around. It is their lack of grip that makes snow and ice slippery. This is a question of academic importance to certain physicists. But here is a fact: the Sami skiers did not care. They cared about reindeer. They learned by trial and error, not through the fundamentals of physics. My son cares no more than a Sami skier. He cares about fun and exuberance and feeling alive on a cool fall afternoon just beneath the treeline. He learns by falling and getting up and falling again.

Next to the trail we see a young moose, nascent antlers still intact, lounging in a snowbank, protected from the wind by a depression in the snow, chewing. Hot clouds puff out of his nostrils, and his jaws move in grinding circles, chewing cud. His expression, while dumb, is not quite so dumb as that of a cow. The peak of Denali—Mount McKinley—is visible far away to the north, beyond Cook Inlet. The mountain, from here, is entirely covered with snow. The weather on Denali at this time of year would be of Martian intensity, but at this distance it looks serene under its shroud of snow.

Someone has stapled a sign to a post next to the trail. The sign is fluorescent red, with the black silhouette of a grizzly bear. In thick letters, DANGER! appears beneath the bear's image, which seems to smile slightly, as if up to something. At the bottom of the sign, in Magic Marker, someone has written, "Moose kill (by bear) near Williwaw Creek. Public use trail closed." It is dated November fifth, a week earlier. We scan the hills hoping that the bear will stand out against the snow. All we see is the shadow of one mountain against another, with the bands of white spruce and shrubby willow creeping up the slopes, the treeline running higher in the gullies, zigzagging its way as far as it can before petering out, and above that just snow, and still higher on the steepest slopes exposed rock, windblown or just too steep to hold snow. It is in some ways like a coral reef, with certain species occurring on the reef flats and others in the gullies protected from waves, and the whole thing petering out in sand. After the Philippines, the wind, though hardly cold by any reasonable standard, bites my cheeks. We ski onward, toward the pass.

※　※　※

Plants, without exception, do not migrate. Some die off each autumn, leaving behind seeds that germinate the following spring, analogous to the overwintering eggs of insects. Some die off above-ground but have roots that survive underground. Certain grasses and herbs and shrubs overwinter with their branches and leaves buried under snow, dormant and hidden, hibernating. Trees, though, stand out in the open, dormant but exposed, not dug into a burrow or cave or snow mound like some cowering, thin-blooded hibernator. The dahurian larch is the world's northernmost tree of any real size. It looks something like a spruce, but its needles turn orange and fall off each autumn. It can be three feet thick at the base of its trunk. It survives the winter cold of the far north, standing in the open at temperatures of ninety below zero, not counting windchill. Its forests look like stands of Christmas trees, sometimes densely packed but sometimes scattered, in Mongolia, North Korea, northern China, and Siberia. It grows as far north as the Khatanga River valley on Russia's Taymyr Peninsula, the northern extent of mainland Asia. The latitude there is seventy-two degrees, two degrees farther north than the treeless tundra of Alaska's North Slope, where the dahurian larch is absent not because it cannot survive, but because the vagaries of geobotanical history and the total absence of landscape gardeners have never brought it there. Make no mistake: this is one tough tree. The dahurian larch is the plant kingdom's answer to Apsley Cherry-Garrard and Father Henry and Ernest Shackleton. It knows how to handle cold weather.

Other trees, softer than the dahurian larch, drop out one by one as the weather becomes colder. First the palm and the mosquito-infested mangrove, killed by nothing more than a hard freeze, and then the shade-tree live oak with its thick branches spread out like umbrellas, dead at eighteen degrees. The redwood and the southern magnolia and the slash pine give up at five degrees above. The sweet gum is gone at seventeen below. Then the maple and the shagbark hickory and the hop hornbeam give way to the frost. And finally it is too much even for the spruce and the dahurian larch, and the forests are gone. Almost gone, because there are still the bonsai trees, the diminutive trees that hardly seem like trees at all, except that they have woody stems and leaves and they are in fact species of birch and willow. There is *Betula nana,* a birch that grows no more than a few feet tall and looks nothing like its full-size cousin the white-barked *Betula papyrifera,* the paper birch, the tree of the birchbark canoe. And the willows: *Salix arctica, Salix ovalifolia, Salix reticulata,* and others, none more than a few inches tall. Compare these to *Salix nigra,* the black willow, the tree of the shifting sand islands and banks of the Mississippi River, its trunk obese, its branches reaching higher than a ten-story building, its leaves longer than the full height of its Arctic cousins, its ability to survive a real winter virtually nonexistent.

Some of the tiny cold-climate willows spread like vines across the ground, but others grow like stunted stately trees, with thick trunks and smaller branches and dense accumulations of tiny leaves, all reduced, all low to the ground, forming forests odd not only for their tiny size but also because the grass, in summer, towers above the trees. An ecologist looking at these miniatures might describe an understory of trees and a canopy of grass. Reduced size lets these willows hide under snow, avoiding the worst of the cold and the moisture-starved winter winds that would suck the water from frozen sap. Without snow cover, the tiny trees would be forced to replace lost moisture, to somehow pull water from frozen ground. More likely, the trees would die.

Many trees — perhaps most — can survive at temperatures far

below those found where the trees grow. Cold of twenty below will not kill the bald cypress, but the bald cypress is not found in climates this cold. The same is true for the Oregon white oak and the red pine and eastern hemlock. Even the paper birch, the quaking aspen, and the black spruce, trees of the far north, would survive in temperatures colder than those where they are found. But it is not only the cold and dry air they have to face. It is the wind, and even the weight of snow. The wind, kicking up ground blizzards and carrying sharp-edged ice crystals, can sand-blast a tree, stripping off bark on the upwind side and exposing raw tissue, killing and polishing the cambium, the tissue beneath the bark that shuttles nutrients between leaves and roots. Blowing snow can kill the upwind branches, leaving a tree that looks like a flag, its living branches all pointing downwind. It can prevent trees from growing upward, leaving them prostrate on the ground. Or, for trees that struggle through the first ten or twenty or thirty feet of height, it can leave a tree mopheaded, able to form thick concentrations of needles on branches high above the blowing snow and ice of ground blizzards.

Accumulated snow, piled on in the absence of wind, can snap a tree in half. Trees forty feet tall sometimes hold more than six thousand pounds of snow. Certain trees have evolved to shed snow. The branches of spruce trees arc gracefully downward, allowing snow to slide to the ground. Fir boughs flex, dumping snow without breaking. When an ice storm hits a forest of oak or maple or ash before leaves have been shed, the weight of ice on leaves fractures branches, limbs, and even trunks, or pulls whole trees over, ripping their roots from the ground, sending them tumbling into houses, laying them across roads, or suspending them in mid-descent on power lines, wreaking havoc. In Finland, on certain mountain slopes, breakage from snow loading controls the extent of forestation. The treeline is controlled by the weight of snow.

A tree that tolerates sand-blasting and survives snow loading still has to withstand freezing. A tree cools gradually, following the air temperature downward. But like the blood of certain insects, the tree's fluid supercools, dropping below the freezing point without actually freezing. At some point, supercooling fails, and the liquid between the tree's cells freezes, not gradually but suddenly. It flash freezes. Molecules, one moment drifting about with the freedom of a liquid state, lock into place, still. In doing so, they dump energy. The act of freezing, of changing from a liquid state to a solid state, releases heat. This is not an abstract concept. Insert a thermometer in a tree, wait while the tree freezes, and record a temperature spike as liquid turns to solid. A beech tree cools steadily to seventeen degrees, the spaces between its cells flash freeze, and its temperature spikes back up to thirty. And then it cools some more, going back into a steady, slow decline.

Outside the cells, in the spaces between them, the tree is frozen. But inside the cells, the fluid is full of dissolved solids, salts, and sugars. The complex molecules of the factory of life float about in each cell's cytoplasm, suspended by water molecules, and the water molecules in the cell are moving around, bouncing off one another, dancing like ten-year-old boys souped up on marshmallows. Some dance right through the cell membrane and, once outside, freeze. This is a form of cryosuction, the same cryosuction that sucks liquid water toward layers of segregation ice in soil. It is cryosuction pulling water molecules across the cell's membrane, drawing them out of the cell. At the same time, the cell membrane opens up, yielding to cryosuction. In cold-hardened trees, cell membranes become increasingly permeable, letting the water go before it can freeze inside the cell and create deadly sharp ice crystals within it. This is part of the reason trees can handle much colder temperatures at the beginning of winter than at the beginning of autumn. As a consequence of cold hardening, the fluid inside cells—the cytoplasm—becomes less wet. The dissolved salts, sugars, and proteins become more concentrated. As with all liquids, the greater the concentration of dissolved solids, the lower the freezing point. This protection is not without limits. At some point, the cells freeze, and ice crystals inside the cells spell death. Or the cells do not freeze, and the concentration of dissolved solids inside the cells increases beyond tolerance, spelling death. Either way, cells die. Kill enough of them, and the tree itself goes the way of Robert Falcon Scott or Lieutenant George De Long or the three frozen sons of Johann Kaufmann in the School Children's Blizzard of 1888.

❄ ❄ ❄

It is November twenty-eighth and minus fifteen degrees at the Chena Hot Springs Resort, sixty miles west of Fairbanks. For ten dollars, I walk on a concrete floor, through a locker room, into a heated indoor pool room, and then outside—in a bathing suit and bare feet. Hitting the minus-fifteen-degree air is like walking into a brick wall. I can feel the cold inside my nose, a feeling of dry boogers that I know are not boogers at all, but ice.

The rock pool, steaming with geothermal heat, is fifteen cold paces away. Steam rising from the pool freezes to the first thing it finds. The rails around the rock pool are covered with ice. Beyond the rails, the pool is surrounded by boulders, also covered with ice. Behind the boulders are trees, what seem to be small spruce trees, covered, too, with ice. The ice covering the rails, the boulders, and the trees is white, not clear.

I wade into waist-deep water and immediately submerge, soaking up the heat. The rock pool is four feet deep and well over a hundred degrees, the temperature of a hot bath. The smell of sulfur rises with the steam. I wade toward the far end of the pool, where the hot spring spills into the pool, and the water is noticeably warmer. The cold air, over a distance of maybe thirty feet, is sucking the heat right out of this water. Gravel covers the bottom of the pool. Green and red lights illuminate the steam coming off the water. The sky beyond the steam is crystal clear, frozen into transparency. Low on the horizon, through the steam, snow-covered hills reflect starlight. Overhead, the stars themselves burn—the Big Dipper and, almost directly above, the North Star. I reach up to run my hand across my scalp. My hair is frozen. I submerge, and the ice, for a moment, disappears.

A man speaks from the edge of the pool. "I'm celebrating my son's house," he tells me. This seems to me an odd statement, out of the blue, to emerge from the steamy shadows at the edge of the pool. I move closer, expecting to see someone with him, but he is talking to me. "We've been working on it for three years," he says, "and today was the final inspection. We're done, and we're celebrating. It has double-insulated walls." The man has lived in Fairbanks his entire life. He is Bergmann's Rule personified, at least three hundred pounds, insulated from the cold of Fairbanks by thick flesh that gives him a rounded appearance, something like that of a northern seal. He tells me that he is in his sixties. He seems eager to explain that he is retired, maybe as an explanation for his need to converse, to feed a hunger for conversation that was once filled by coworkers. But it is not so much conversation as monologue. It is not clear to me that my presence actually matters. Increasingly, I feel like one of the monkeys at Jigokudani Park in Japan, which like to bathe in hot springs through the Japanese winter, jabbering in meaningless primate patter, surrounded by steam and frost, falling into a stupor induced by heat and sulfurous fumes.

"We had snow in June this year," the man says. "Snow and a hard frost. It killed all of the flowers in the yard. Of course, we had more flowers. The garage was full of flowerpots. Only the ones in the yard were killed. It slowed down the construction season, too. We got a late start on my son's house. But it's done now."

I dunk to thaw my hair. Is it rude to dunk in mid-monologue, the only listener suddenly disappearing underwater? But when I surface, he is still talking. "There's not enough snow to ski," he is saying. "Last year we had better snow. When I ski, I ask for the slowest skis they have. I'm scared of skiing. I'm afraid of skiing too fast." He is too obese for skis. It is hard to imagine him on skis. My hair is frozen again, so I submerge and inch backward. When I resurface, he is hidden by steam, but his voice goes on as if I have not moved. I continue to back away until his voice is muffled in steamy darkness.

Later, leaving the rock pool, I find my towel stiff as a board, coated with frozen steam. Before I get inside, the hairs on my arms and chest freeze. They become brittle. When I move my arms, I can feel the hairs breaking under the strain. I am pleasantly overheated from the pool. I feel slightly dizzy, the stupor of a Jigokudani Park monkey. Eight inches of ice hang from the handle of the door going into the heated pool room. Just outside the locker room, a map with pushpins shows the homelands of visitors. There are three pins in Japan, one in Afghanistan, many in China, one in Belarus, a few in Russia. The Lower 48 states are well pinned.

Dressed, I walk for a few minutes in the compound. My hair, still damp, freezes again. My ears grow cold, but I am otherwise warm, parboiled, wearing only a sweater and a light jacket. On one side of the compound, the owners have erected an entire building made from ice. They call it the Aurora Ice Museum, but it looks more like a church. Inside, there are ice sculptures. There is an ice bar, with fifteen-dollar martinis served in glasses made from ice. It is possible to rent the church of ice for corporate events, for birthday parties, for weddings. Wedding ceremonies are held at an ice altar. For just under two thousand dollars, guests can participate in a three-day ice-sculpting class.

Elsewhere, for around six hundred dollars, it is possible to stay in an ice hotel. Ice hotels are a new phenomenon, the first one appearing in 1989. Most winters, tourists have choices. There are at least six solid-ice hotels in the world, in places such as Norway, Sweden, Finland, Romania, and Quebec.

By the time I get to my traditionally built hotel room, I have cooled off. The night has dropped another three degrees, to minus eighteen. In my room, a sign taped to the air conditioner says, "Please do not use air conditioning September 15th to May 15th."

❄   ❄   ❄

Charles Darwin, seasick, sailed into Tierra del Fuego, near the southern tip of South America, in 1832. It was December, the middle of summer in the Southern Hemisphere. "The climate is certainly wretched," Darwin wrote. "The summer solstice was now passed, yet every day snow fell on the hills, and in the valleys there was rain, accompanied by sleet. The thermometer generally stood about 45 degs., but in the night fell to 38 or 40 degs."

He described the forests, noting in particular the treeline: "The mountain sides, except on the exposed western coast, are covered from the water's edge upwards by one great forest. The trees reach to an elevation of between 1000 and 1500 feet, and are succeeded by a band of peat, with minute alpine plants; and this again is succeeded by the line of perpetual snow."

The trees survive as best they can, but it was not so much the trees that interested him as the people.

> While going one day on shore near Wollaston Island, we pulled alongside a canoe with six Fuegians. These were the most abject and miserable creatures I anywhere beheld. On the east coast the natives, as we have seen, have guanaco cloaks, and on the west they possess seal-skins. Amongst these central tribes the men generally have an otter-skin, or some small scrap about as large as a pocket-handkerchief, which is barely sufficient to cover their backs as low down as their loins. It is laced across the breast by strings, and according as the wind blows, it is shifted from side to side. But these Fuegians in the canoe were quite naked, and even one full-grown woman was absolutely so. It was raining heavily, and

the fresh water, together with the spray, trickled down her body. In another harbour not far distant, a woman, who was suckling a recently-born child, came one day alongside the vessel, and remained there out of mere curiosity, whilst the sleet fell and thawed on her naked bosom, and on the skin of her naked baby!

Since Darwin's time, the Fuegians—the Yamana, Selk'nam, Manek'enk, and Alacalufe—have been admired for their ability to withstand the cold. Take a man of European, African, or Asian heritage, lay him down next to an Alacalufe native of Tierra del Fuego, and it is not the native who is abject and miserable, but the shivering European or African or Asian. The European or African or Asian shivers to maintain a reasonable body temperature, while the Alacalufe stays warm without shivering, through what is sometimes called "nonshivering thermogenesis." They have physiologically adapted to cold, with a metabolic rate as much as forty percent higher than that of other races allowing them to maintain a normal body temperature while sleet runs off their skin.

In Darwin's day, the Aborigines of Australia slept naked on the ground, even in the colder southern regions of the continent, where temperatures might drop below freezing. Unlike the Alacalufe, the Aborigines did not stay warm. They had adapted to cold through a different path than that taken by the Alacalufe. An Aborigine lying on the ground would become colder than a European, an African, or an Asian. His body temperature would drop. When it hit the point that would trigger shivering in the European or African or Asian, the Aborigine would not shiver. His body temperature would keep dropping. He would enter a state of shallow hypothermia, unperturbed. During the night, his core temperature might drop four degrees, to ninety-five degrees. He would then enter what has been called a state of "nightly torpor," perhaps something like that of hungry titmice or finches, which in turn is something like daily hibernation, which in turn has been compared to suspended animation. Today Australian Aborigines sleep in heated homes, but presumably they could still enter into nightly torpor, maybe dreaming through shallow hypothermia, while other Australians would shiver, likely not sleeping at all, and would appear, and in fact be, abject and miserable.

There may be differences, too, among Europeans and Africans and Asians. During the Korean War, it became evident that frostbite was more prevalent among black soldiers than among white soldiers. On average, black men were four times more likely than white men to suffer injuries from the cold, and black women were twice as likely to suffer injuries as white women. "Arabs," wrote a doctor reporting these data, "appear to be similarly predisposed, as are individuals from warmer climates." But individual variability would be considerable, with some black men and Arabs facing the cold with far more aplomb than most of their white colleagues.

Blood circulation to the skin and hands is greater in Inuit people — Eskimos — than in Europeans, apparently protecting them from frostbite. Norwegian fishermen have shown similar tendencies. Korean and Japanese ama divers — traditionally women who dive on a breath of air after seaweed, shellfish, and pearls — have a legendary tolerance for cold water. A century ago, they often dove topless, spending hours each day in the water. Today they wear wet-suit tops or tights. Whatever physiological tolerance they once had has diminished but not disappeared.

A certain amount of adaptation is possible. In 1960, a University of Alaska physiologist named Laurence Irving noticed two students wandering around the Fairbanks campus in light clothes and without shoes. The students were adhering to the rites of a religious sect that discouraged the wearing of shoes, a practice that must have originated somewhere warmer than central Alaska. Irving put the students in a room chilled to thirty-two degrees, then measured the temperature of their fingers and toes for an hour. Their toes and fingers would cool off, then warm up again, apparently as blood vessels constricted but then dilated, sending warm blood to the extremities before they became dangerously cold. The students did not shiver until after fifty minutes. Irving did the same with an air force volunteer but had to discontinue the experiment after thirty minutes. "The airman's toes," he wrote, "became so painful and he began to shiver so violently that I caused him to terminate the test lest he shake himself apart."

Quebec City postal workers, walking from mailbox to mailbox through the Canadian winter, grow more cold tolerant as winter wears on. Heart rate and blood pressure drop. Workers in Antarctica are said to adapt over time by increasing their core temperature. Charles Wright, who traveled with Scott in Antarctica but was not chosen for the fatal trek to the pole, trained himself for the cold even as a child. "For some incredible reason," he told an interviewer, "I thought it was a good thing — I was living in Toronto at the time — to toughen oneself a bit, so I wore the same clothes in summer and winter."

Habituation to cold might disappear after a few winters in Florida but be regained with a return to colder regions. By contrast, the genetically programmed tolerance of the Alacalufe, the Australian Aborigine, and the Inuit would be more long lasting. But all of this habituation in human postal workers and Arctic explorers, all of the genetically conferred advantages of the Alacalufe and the Australian Aborigine and the Inuit, offer no more than the smallest advantages. Compared to the arctic fox or the wolf or the musk ox, compared to the ground squirrel or the wood frog or the poorwill, the most cold-hardened human, if forced to rely on physiology alone, is as fragile as thin ice. A human without thick clothes or shelter or fuel will freeze to death in conditions too warm to trigger shivering in a moose. At the end of summer, the human response is primarily one of putting on clothes, lighting a fire, and turning up the thermostat. Or, alternatively, of frostbite and death by hypothermia.

DECEMBER

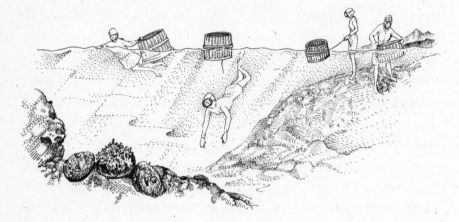

It is December third and twenty-six degrees in Anchorage. Heat leaks through roofs, and melting snow drips over gutters, turning into thick icicles that threaten to crash down, dangerous, not sharp enough to impale but heavy enough to bludgeon. The sky is overcast, the clouds forming a vaporous blanket that holds in the heat. Thanksgiving week, with six straight days below zero, has defrosted. The clear, cold nights of November, loitering in the teens, have migrated.

The entire state suffers in the heat. Juneau, the capital, hit thirty-nine. Anchorage is warmer than Chicago, Des Moines, Minneapolis, Salt Lake City, and Grand Rapids, and almost as warm as Omaha, Seattle, and Albuquerque. The North Slope hit twelve degrees. Fairbanks, in the icy interior, is in the positive digits, four above.

This is inconvenient but not terribly unusual. In 1929, Anchorage hit fifty-one degrees on December third. People in Anchorage grow accustomed to winter heat waves driven by what fur traders called the chinook winds — winds that came from the land of the Chinook tribe, in the Pacific Northwest. The Alaskan version of the chinook winds blow into Anchorage with some regularity, their hot breath melting the snow, leaving streets filled with water and turning forest floors into slushy swamps. It happens so abruptly that the chinook winds are sometimes said to eat the snow. On the mountainsides above Anchorage, chinook winds can reach hurricane strength. The loss of roofs from hillside houses is not unknown, giving wealthy homeowners exceptional but unexpected views of crisp winter skies. Anchorage is not unique in suffering from these winds. In 1972, a chinook wind blew into Loma, Montana, raising temperatures from 54 below to 49 above, a change of 103 degrees in twenty-four hours.

But today's heat wave did not blow in on a chinook wind. It was forecast by the National Weather Service in October. Using data from satellites and ocean buoys moored in water more than three miles deep, the National Weather Service warned that we would have an El Niño year. A week ago, the Climate Prediction Center in Maryland issued an advisory: "El Niño conditions should intensify during the next one to three months." For Alaska, this December through February will be warmer than normal.

Long before it came to the attention of meteorologists, El Niño was noticed by South American fishermen, who recognized warm currents off Peru and Ecuador each year around Christmas. The warm currents killed the fishing. The fishermen named the annual event El Niño, "the Little One," in honor of the birth of Christ. Globally, El Niño has come to refer to those years when the Christmas currents are unusually strong. The currents strengthen when trade winds blowing to the west weaken and unusual volumes of warm water reach Peru and Ecuador. Global weather patterns respond. It rains in California. Australia dries out, and bush fires burn out of control. Gulf of Mexico hurricanes become less prevalent. South China Sea typhoons become more prevalent. Corals die in the Pacific Ocean. People catch marlin off the coast of Washington State. With El Niño, cold, clear high-pressure air stalls between Alaska and Seattle. The pressure ridge forces warm winds forming along the Aleutian Islands to swing north, toward Anchorage. The city will have rain within days.

This is no chinook wind, but our snow will be eaten all the same.

❄   ❄   ❄

The moose is so well insulated that the southern border of its range is sometimes said to be set by its intolerance of heat. A moose has two kinds of fur—thick guard hairs that give it color and finer underfur that keeps it warm. Piloerection is the fluffing out of hair, often for the purpose of staying warm—for increasing the insulative effectiveness of hair. In humans, piloerection, rendered useless by millions of years of evolution, causes goose bumps. In moose, the fur is so efficient that piloerection does not occur until the mercury drops below minus ten. A moose's metabolism does not increase until minus twenty. But even on a mild winter night without wind, the moose will lose the energy equivalent of a Snickers bar every hour. It needs to eat or lose weight. In fact, it does both.

A moose in Alaska's Denali National Park might feed six hours each day, swallowing more than thirty pounds of branches and bark. Then it ruminates for another twelve hours. In this context, ruminating is not a matter of turning something over in the mind, as a human might do, but rather of chewing cud. Like a cow, the moose regurgitates partly digested food from its first stomach, chews it, and reswallows. The jaw muscles, ruminating, generate heat. The four stomachs, churning, generate heat. Microbial action in the gut generates heat. A ruminating moose might generate fifty percent more heat than a fasting moose. But in winter, the food is tough to digest. Branches and bark are less nutritious than the fresh green shoots and leaves and flowers of spring and summer. The amount of food that the moose can process is limited by the number of hours in a day. It is possible for the moose to starve to death with its stomachs full, unable to keep up with energy demands.

The moose has to decide how hard it will search for food. Though seemingly dim-witted, it knows at some level that it is not likely to find high-quality forage in the snow. It knows, too, that movement through snow is no easy task. It lifts its long legs high to step over the snow and even plows through deep snow, looking for thinner patches, following the tracks of other moose or wind-scoured frozen rivers or ski trails. But when the snow is deep, the moose must reduce its range. It may wander over no more than ten or twenty acres, conserving energy, chewing its cud, burning its fat, stripping bark from trees. It will sit in the snow in preference to standing, using the snow as a blanket. Foraging, successful or not, will use at least twenty percent more energy than sitting in the snow. On sunny days, the moose will move into the open. On cold nights, it will sleep under thick spruce cover. It will burn ten thousand calories or more in a day, twice that of a well-fed Arctic explorer. Like the Arctic explorer, the moose will lose weight.

The moose has been called a "confronter" or a "tolerator" or, more descriptively, a "winter active," because it confronts winter. It tolerates winter. Winter actives neither hibernate nor migrate. The formula they pursue seems simple: take in as many calories as you can, preserve your fat, and hope for the best.

Winter actives have fur that holds in heat. The arctic fox, curled up in a ball, wrapped in its own fur, lies comfortably on the northern pack ice at forty below, steadfastly refusing to shiver. The winter coat of the caribou is so warm that the animal uses less energy in winter than in summer. Musk oxen in small herds of ten or fifteen, when harassed, circle up, horns out and heads down, their hair hanging matted, windblown and iced, their expressions Pleistocene. They refuse to run not because they are stupid, but because their coats are so warm that to run is to overheat. Other animals have layers of fat that not only buffer winter food shortages but also provide insulation. They wear sweaters of fat. An inch of fat provides more insulation than an inch of wool. A male polar bear ambles across the pack ice throughout winter, feeding on seals. It enters the cold season rotund, wearing the equivalent of eight or ten wool sweaters under its fur.

Where fur and fat are not enough, there is shelter. Gray squirrels, unlike ground squirrels, do not hibernate. Instead, they make nests of twigs, maybe ten inches in diameter. Inside, a well-made gray squirrel's nest might have twenty layers of leaves, and inside the leaves might be finely shredded bark, and inside the finely shredded bark might be a four-inch-wide chamber, a gray squirrel's winter parlor, its home between bouts of foraging. Flying squirrels make similar nests. The insulation of a flying squirrel nest can weigh five times more than the squirrel itself. Naturalist Bernd Heinrich once chased a flying squirrel from its nest, put a hot potato in the insulated chamber, and watched it cool. At nine degrees outside, the potato cooled less than thirty degrees in half an hour. Put another way, the potato was still hot enough to eat after thirty minutes in a flying squirrel's nest with the outside temperature at nine degrees above zero.

Winter actives have other tricks and adaptations. Beavers and muskrats huddle with their families, sharing their warmth. The feet of snowshoe hares are so enlarged relative to their weight that they walk on the snow's surface without breaking through. Their foot loads, as they are called, are comparable to those of a human wearing snowshoes that are ten times the size of a human foot. Caribou plant their front feet at a steep angle, supporting their weight in the snow on the hoof, the upper part of each foot, and the dewclaw—a toelike protuberance sticking out from the back of what looks like the animal's ankle. Likewise, lynx, wolves, and wolverines float across the snow's surface on enlarged feet.

Some winter actives store food. The gray squirrel stores nuts. The fox caches eggs and frozen carcasses and biscuits stolen from Dumpsters. The diminutive half-rabbit, half-hamster, half-pint pika—known variously as the rock rabbit, the whistling hare, and the coney—dries herbs in the summer sun, making herb hay while the sun shines, then stores it in the caves and crevices of rock falls. By winter, an individual pika, weighing a third of a pound, may have half a dozen two-pound piles of herb hay scattered in the rocks.

The snow itself is shelter. The subnivean dwellers—the lemmings and voles and shrews—are winter actives. Their tunnels in the snow hover right around freezing through the winter. It is not just a matter of insulation. If it grows colder above their snow tunnels, liquid water in the snowpack freezes, and in so doing releases heat, keeping the temperature in the subnivean caverns warmer than it might otherwise be. If it grows warmer above the snow, the snow goes from a solid to a liquid, and in so doing absorbs the heat. The end result is stability, in some ways as important as warmth. For the subniveans, it is safer under the snow, and through most of the winter, it remains warmer beneath the snow than above, and there is no wind. If winter behaves well, if the chinook winds are not too severe, if unexpected warming does not ruin the winter season, the subnivean winter actives know what to expect. But if it grows too warm for too long, tunnels can flood, forcing the subniveans to the surface, where they face not only the elements but also hungry predators.

The fact is that some of the winter-active animals do more than tolerate the seasonal cold. Some thrive. Lemmings, for example, are known to mate and reproduce in their subnivean tunnels. They have snowbound orgies and fill their tunnels with young, thumbing their noses at winter. The young will themselves be old enough to breed and reproduce and raise their own offspring within weeks of being born. Some will be born and reach adulthood before the spring thaw, when they will see for the first time the light of the sun and feel for the first time the warmth of a summer day.

And there is food in the snow, food for lemmings and voles and shrews. There is a food web here in these subnivean chambers that humans seldom suspect. Between the ground and the snow, through the winter, there are fly larvae, beetles, millipedes, fungi, and bacteria, all of them working over the dead leaves, roots, and branches of the previous summer. Spiders and centipedes and bigger beetles eat the fly larvae and millipedes and smaller beetles. The shrews and lemmings eat anything big enough to draw their attention. This is no desert. This is not some cave with only one or two species. In Canada, a scientist poking around in the subnivean world found nineteen species of spiders, fifteen species of mites, sixty-two species of beetles, sixteen species of springtails, thirty-two species of ants and wasps, and two species of centipedes, all of them active, living under winter snow.

And there are seeds. From William Pruitt's *Wild Harmony: The Cycle of Life in the Northern Forest*:

The new snow covered the layer of birch seeds and hid them from the small birds. The added weight of the fresh snow compacted the middle layers of the *api* [snow cover]. As the crystals squeezed together and broke, the cover creaked and groaned. A foraging vole would stop and huddle, ears twitching, and then resume its errand. The sounds breached the even tenor of life under the snow. An additional disturbance was the faint scent of birch carbohydrate that occasionally filtered down from above. Some voles dug upward through the layers of snow. One layer was easily tunneled, the next was harder; no two were alike. When a vole reached the seed-rich layer, it drove a horizontal drift along it and devoured every seed.

The tunnels are not without danger. They can freeze over with ice crust, becoming impenetrable. Carbon dioxide can build up. When warm snaps and chinook winds melt the snow and send floodwater into the tunnels, voles and lemmings and shrews can drown in their chambers. And if there is not enough snow, if the world grows cold before the snow grows thick, as often happens, the subniveans freeze to death, or starve trying to stay warm, or, as is closer to the truth, starve and freeze to death and succumb to disease simultaneously, like Greely's men and De Long's men and Bering's men.

The interactions between animals and their environment do not flow in a single direction. As the first line of a poem affects the last and the last line affects the first, the environment affects animals and animals affect the environment. Subnivean winter actives change the nature of the snow, creating caverns for meltwater in spring. Above the snow, foraging hares and caribou and moose change the nature of their forage. Along rivers, hares eat willows, and on higher ground they eat birch saplings. This winter feeding can be intense enough to control the structure of the forest, as if the animals are farming the trees, keeping them small and as succulent as possible year after year. But the plants fight back. Invisibly, the plants produce compounds that deter winter grazers. The paper birch makes papyriferic acid. In succulent juvenile plants — plants that appear juicy and tasty — the acid may be twenty-five times more concentrated than in adult trees. At times, it forms droplets on the winter twigs of saplings. Captive hares, offered birch twigs with naturally high concentrations of papyriferic acid, stopped eating. There is reason to believe that the coming and going of some hare populations, usually blamed on hungry lynx, may be a result of chemical defenses in plants. As the hares graze, the plants produce more of the compounds that deter grazing, becoming less edible, and the hares die back. With fewer hares and less grazing, the plants become more edible, and the hares grow more abundant.

Winter for hibernators is safe; for bears, as few as one percent might die in their dens. Winter for winter actives is dangerous. It is not the single cold spell that kills a caribou or a moose, but rather the additive effect of cold spells and snowstorms and missed meals, or even the effects of one winter adding to the next, with summers too short for the animal to recover fully. One summer, they are a bit on the slim side, the next summer they have slimmed down a bit more and do not reproduce, and that winter they die from starvation or disease or, too slow to escape, from a set of teeth ripping through the jugular or crushing the skull. A pack of wolves can consume a moose every few days. The wolves will also eat caribou, beavers, hares, voles, and shrews. Shrews are also subject to attacks by weasels and owls. While a shrew hunts bugs under the snow, an owl or a weasel or a fox hears the commotion and crashes through from above, bringing sudden daylight and death into the tunnels below.

Winter actives must deal with humans, too. Snowmobiles and skiers spook the animals, pushing them into an energy-wasting run. Fences prevent them from reaching better grazing. Roads make convenient corridors for winter movement, luring animals to death by unexpected impact. In Wyoming, in 2004, a vehicle ran through a herd of pronghorn, killing 17 animals. In 2006, it happened again, another vehicle in a different place killing, coincidentally, another 17 animals. Railroads, too, make deadly corridors. A train heading west through Wyoming one winter ran through a herd of winter-active pronghorn, rendering 125 of them suddenly and permanently inactive, dead on the tracks.

❄ ❄ ❄

It is December tenth and thirty-seven degrees in Whittier, Alaska. Six inches of slush cover the parking lot, and rain falls at a slant, carried by wind. I am wearing a dry suit, a scuba tank, and thirty pounds of lead. Under my dry suit, I wear two nylon T-shirts, a wool sweater, sweatpants, thick socks, and coveralls made of Thinsulate. Three-finger mitts, a quarter of an inch thick, cover my hands. The water is thirty-nine degrees, two degrees warmer than the air. A sea otter swims on the surface, watching me dive. Below, stunted kelp dusted with glacial sediments grows on rocks and gravel. On the seabed, spot shrimp dart about between rocks. Multiarmed starfish crawl around on their tube feet searching for shellfish. At eighty feet, I swim through a field of pale sea whips, a cold-water soft coral that grows in vertical ropes rising six feet from the bottom. The stalks are covered with tiny eight-armed polyps. Nearby, two copper-colored rockfish sit on the bottom, thick finned, huddled in depressions in the mud, as if trying to stay warm. The kelp, the shrimp, the starfish, the corals, and the fish all have enzymes that work reasonably well in thirty-nine-degree water. I do not.

Dry suits always leak. They are not dry suits so much as damp suits, with water sneaking in through the wrist and neck seals. Moisture leaves my arms and chest clammy. At twenty minutes, my hands are numb. The blood vessels in my fingers, especially near the skin, have squeezed shut, dropping the blood flow to ten percent of normal, conserving heat for my core. I can still use my mitted hands to work my dry suit valves, but it is increasingly hard to move individual fingers. My air consumption increases. At forty minutes, my hypothalamus, buried deep in my brain, somewhere behind my nose, triggers shivering. As a rule of thumb, shivering starts when the core temperature drops to ninety-seven degrees, two degrees below normal. Muscles contract and relax in an effort to generate heat, cycling six to twelve times every second, burning glycogen like there is no tomorrow, generating four times more heat than the body at rest. When the glycogen is gone, the shivering stops. Or if the body temperature drops to about eighty-eight degrees, the shivering stops, and then, in all likelihood, there is no tomorrow.

I try to think warm thoughts. I repeat the mantra "I am warm. I am warm. I am warm." In fact, I am not. The mantra fades, and

I think of Laurence Irving's airman, experimentally exposed to the cold, shivering so violently that Irving worried that he might shake himself apart. I am abject and miserable, wishing that I had the blood of one of Darwin's Fuegians. I think of my caterpillars, Fram and Bedford, lying curled up in my freezer, frozen solid. I envision a ground squirrel in its hibernaculum, shivering when its body temperature drops below freezing, warming up and then drifting back into the stupor of cold in a cycle that repeats itself through the winter. And then there is Apsley Cherry-Garrard, writing of the relative warmth of fifty-five degrees below zero.

The *U.S. Navy Diving Manual* has this to say:

> Hypothermia is easily diagnosed. The hypothermic diver loses muscle strength, the ability to concentrate and may become irrational or confused. The victim may shiver violently, or, with severe hypothermia, shivering may be replaced by muscle rigidity. Profound hypothermia may so depress the heartbeat and respiration that the victim appears dead. However, a diver should not be considered dead until the diver has been rewarmed and all resuscitation attempts have proven to be unsuccessful.

For navy literature, these might pass as reassuring words: you are not dead until you are warm and dead.

The manual also warns that regulators can freeze underwater, dumping the diver's air supply in a steady free flow. And it says, in one understated sentence, that "hypothermia may predispose the diver to decompression sickness." At any temperature, under pressure, nitrogen in the diver's air supply dissolves in the blood. If the diver surfaces too quickly, the nitrogen leaves the blood as bubbles, damaging tissue and blocking capillaries. The results can mirror those of a stroke: pain, numbness, paralysis, loss of memory, dizziness, death. But low temperatures increase the solubility of nitrogen, so the cold diver takes on nitrogen more quickly. Worse, as the diver grows colder, constricted blood vessels in the hands and feet do not allow the blood flow needed to flush dissolved nitrogen efficiently. Worse still, it quickly becomes too cold to seriously consider a slow ascent or a decompression stop. By now, though, I am too cold to focus on any of this. The thoughts just pass through my mind, as if part of a dream. I am dangerously cold to be underwater. I signal my dive partner that it is time to head up.

I am soft. The ama divers of Japan and Korea would find my softness amusing. For more than a thousand years, they have picked up seafood, diving on a breath of air. For a brief period, they dove with helmets and canvas suits, but they saw that this sort of diving would soon exhaust the fishery. Rules emerged that restricted divers to a breath of air. In most places, an ama can wear nothing warmer than a partial wet suit. The thinking is that only the toughest and most skillful divers should succeed. They work in water as cold as fifty-seven degrees. At this temperature, softer people are exhausted or unconscious within two hours, and predicted survival times are less than six hours. The ama often work for three or four hours at a time. Almost all of them are women. Men, it is said, cannot handle the cold. Some of the ama dive into their seventies. They are not demure. They have a reputation of independence, of rude language, of loud voices. An anthropologist who once lived and dove with the ama remarked on their swearing and summed them up by describing their means of greeting one another: they would say, in Japanese, "Yo!" instead of "Hello." The ama, she wrote, stay in the water as long as they can, becoming at least mildly hypothermic on a daily basis. They return to shore and huddle inside their *amagoya* warm-up shed. The anthropologist wrote of returning to the *amagoya* one afternoon. The ama were withdrawn and sullen after an unsuccessful day in the water. It was raining. Everyone was cold.

As they warmed up, they began to laugh and joke. Two of them put on costumes to entertain the others. "You see," one of them told the anthropologist, "if we laugh we can forget how damn cold it is!"

They would laugh at me now, with all my gear and still shivering, a weak American male. If I become much colder, it will be difficult to hold this regulator in my mouth. When I reach the surface, the first thing I see is a small waterfall, flowing over the edge of a bluff and falling twenty feet into the ocean. The water flows from beneath a sheet of melting ice surrounded by spruce trees, their boughs heavy with slushy, dripping snow.

❄ ❄ ❄

In the water, large whales are less like me and more like moose and musk oxen: they are as susceptible to overheating as they are to hypothermia. Water sucks away heat thirty times more quickly than air, but the large whales wear a coat of blubber, far more than a sweater of fat, more like a down coat without zippers, permanently affixed. A matrix of collagen gives blubber a spongy structure that yields to pressure but does not sag or jiggle like the fat of obesity. The stuff has the thermal conductivity of asbestos. In mammals with thick coats of fur, the temperature rises between the outside of the coat and the skin. But the skin of a whale is close to the temperature of the water in which the animal swims, with the temperature rising only within the blubber, closer to the inner furnace of the whale. In the bowhead whale — a sixty-foot-long baleen whale of the far north — the blubber can be more than two feet thick. Outside, when the whale swims between and beneath ice floes, and when it surfaces to breathe through cracks in the ice, its skin temperature could be close to freezing, but inside the temperature holds close to ninety-six degrees. And the blubber is vascularized, with shunts that control the amount of blood reaching the skin. The shunts open when the whale gets hot and close when it gets cold.

The blubber is more than insulation. It streamlines the body, and near the dorsal fins and in the tail the blubber may act as a biomechanical spring, storing and releasing the energy needed to flap huge flukes through thick water. The blubber floats the whale, like an enveloping inner tube, like a diver's buoyancy compensator.

The bowhead whale's food generates the heat that the blubber protects. The whale swims through the water with its mouth agape, or skims the surface, or occasionally takes a mouthful of mud. With its tongue, it squeezes the mouthful of liquid oozing with mud and crustaceans and larval fish outward through the baleen combs that look like black, feathery plastic teeth. This is not just any tongue, but rather a tongue of tongues, fifteen feet long and ten feet wide. The tongue squeezes out the water and then sweeps across the bristles of keratinous baleen, the combs that pass for a sieve of teeth lining the whale's giant mouth, combs that may remind one of teeth but that in fact evolved from the ridges commonly found running along the roof of the mammalian mouth, including the human mouth.

The bowhead, despite its blubber, loses something like ten thousand calories each day to the cold. The crustaceans and fish and mud that the whale scoops from the sea are not pure fat. To get the fat it needs, the bowhead requires a hundred tons of food each year to stay warm, plus that much again for growth, movement, and the making of little whales. To complicate matters, the bowhead feeds for the most part during the summer and autumn, when it may consume as much as two tons of food in a day, then diets through the winter and early spring, when food is less abundant.

For bowhead whales, the cold is not just the cold—it is ice. Swimming beneath the ice is no small risk for an air breather. At times, the bowhead dives between openings in the ice, holding its breath, perhaps following bubble trails left by predecessors, or looking for light penetrating down from above, or following the low-pitched rumbling calls of its brothers and sisters and parents and cousins as it makes its way sometimes for more than a mile between surfacings. At other times, it follows leads in the ice, long openings between two ice floes, that, as any Arctic explorer will tell you, can close with little warning. Inupiat hunters say they have seen bowheads break through ice two feet thick. Scientists, who in comparison to Inupiat hunters are laughably ignorant of the way of the whale and spend far too little time on the ice to see what is really going on, have documented bowheads bashing through ice seven inches thick. And the whales have another trick: they can find a tiny airhole and then push the ice upward, not breaking it but merely forming a hummock, a bump on the surface, into which air is sucked and from which the animal breathes.

Also swimming under the ice are the much smaller narwhal and beluga—the tusked whale and the small white whale of the north, both toothed whales more akin to dolphins than to the mighty crustacean-and-mud-eating baleen whales. The narwhal has been found within three hundred miles of the North Pole, and the beluga within six hundred miles, relying on open leads in the ice for air.

There are seals, too, out on the ice. The ringed seal is found as far north as the North Pole, living in winter beneath and in the ice, using its foreclaws to scrape away breathing holes, maintaining bigger openings that lead to snow caves where the animals give birth and nurse their young. In spring, they climb out on the ice to bask in the sun and to shed their fur and replace it with new fur. They have blubber, but they also rely on fur for warmth. The individual hairs of seal fur are flat, not round as they are in most carnivores. The flatness lets the hair lie down, streamlining the animal as it swims.

And there is the sea otter—blubberless, but with more hair per square inch than any other animal alive. Shave a square inch of sea otter, and you will have to sweep up nearly a million hairs. Its hairs trap a layer of air, and the otter, though living in water, is never quite wet. It wears a dry suit that actually stays dry. In exchange, it spends more than one hour in ten preening, combing, and grooming. It sometimes lies on the surface, belly up, its nose and paws—unprotected by the fur that covers its body—held out of the water for warmth. And like whales and seals and birds and to some degree even humans, the sea otter isolates the warm blood of its core from the cold blood of its extremities with a rete mirabile—literally, a "wonderful net," a mesh of veins and arteries. The mesh is a heat exchanger. Arteries carrying warm blood from the core are surrounded and interwoven with veins carrying cold blood from the extremities. Before the warm blood hits the extremities and gives up its heat to the outside world, it warms the cold blood that is moving back from the extremities toward the core. Heat is conserved. Calories are saved.

Often whales and seals and otters are the hottest things around. A Weddell seal, a thousand pounds of fur and blubber and heart and lung and rete mirabile, might lie on the Antarctic ice, open the shunts that let warm blood flow through its blubber, and create above it a cloud of steam. After a time, bored or hungry or spooked by a nosy human, it might flop from the ice into the water. It might leave behind the marine mammal equivalent of a snow angel, an outline of itself melted into the ice, a negative image of belly and fins and head in three dimensions. The Weddell seal thumbs its nose at the cold, leaving in the ice an image that is often called a seal shadow.

❄   ❄   ❄

It is December sixteenth, and the mercury hovers around zero near Lynx Lake, ninety miles north of Anchorage. Wind from the north has broken the heat wave, sending the El Niño warmth scurrying south. We ski across one frozen lake after another: Ardaw, Jacknife, Bald, Frazier, Little Frazier, Lynx. There are three of us: myself, my son, and a companion, the same woman who joined me on Ben Nevis in September, the woman with Raynaud's disease. Between the lakes, we ski across spits of snow-covered land, what would be canoe portages during the summer. The snow is thin. In places, we ski on patches of exposed ice. We pass a place where my son and I camped in August, huddled in a tent in the rain next to our canoe. Some of the lakes are joined by frozen creeks that run through frosted peat bogs and iced-over marshes. Marsh plants sticking above the snow have sprouted ice flowers—blossoms of white ice poking through dry brown stems. The stems, just before they freeze, hold liquid water. As the water freezes, it expands, bursting out of the stems, forming these blossoms. The freezing water draws liquid water behind it, and the blossoms grow. The blossoms here are the size of Ping-Pong balls, but I have heard of ice flowers as large as baseballs.

We cross a set of moose tracks, a beaver lodge, fox tracks. I watch and listen for birds but see and hear none. The air is still. If we stop, the world seems muffled. Sound is nothing more than changes in pressure. Pound a drum, and the air beneath the drum is compressed and released. Speak, and the vocal cords compress air in the larynx. The compressed air moves outward in a wave. As sound waves travel past snow, they momentarily increase the air pressure, forcing air into pores between snowflakes and ice crystals. As the pressure of the sound waves drops, the air moves out of the pores. The air is moving in and out a hundred times per second at a low pitch, five thousand times a second at a medium pitch, and eighteen thousand times a second at a high pitch, near the edge of human hearing. With each movement in and out of the snow, energy is lost. The snow has swallowed the sound. When we move, our skis drag across the snow, torturing it, making it scream with every gliding motion. At warmer temperatures, it is more of a crunching sound, but near zero the pitch increases. It becomes more akin to the sound of fingernails on a chalkboard. The crystal structure is stronger. What we hear is the sound of ice crystals being crushed and torn. This screeching is the sound of hydrogen bonds unbonding under the weight and movement of our skis.

Between two lakes, we ski on a creek. The ice sounds hollow. In places, the snow has blown away, and I can see that we are skiing on three inches of ice, under which there is six inches of air, and under the air, more ice. When the creek was freezing, the water level was dropping. The space between is more than big enough for muskrats.

I worry as we move toward an unseen cabin. Earlier, I had trouble with the car. I had forgotten certain gear. I had expected warmer temperatures. And I had hoped for more snow. If a breeze comes up, these cold temperatures will turn brutal. There will be windchill, but also blowing snow, the thin layers of it lifted off the ice, sandblasting exposed skin. I worry about chilblains, the nasty blistering and sores that can erupt when fine blood vessels constrict in the cold and leak blood beneath the skin. I worry about frostnip and frostbite. I worry about hypothermia and death by exposure. It was not far from here that Andrew Piekarski died under his lawn mower, pinned to the ground while he slowly froze to death, way back in September.

And I have in my mind something J. Michael Yates, a poet and playwright, wrote of a man who "has been moving north always.... On his back he carries pack, snowshoes, and rifle." The man dies:

A man, warmly dressed, in perfect health, mushing his dogs a short distance between two villages, never arrives. He has forgotten to reach down, catch a little snow in his mitten, and allow it to melt in his mouth.

For a reason neither he nor his dogs understand, he steps from the runners of his sled, wanders dreamily—perhaps warmly, pleasantly—through the wide winter, then sits to contemplate his vision, then sleeps.

The dogs tow an empty sled on to the place at one of the two villages where they're usually fed.

While those who find the frozen man suspect the circumstances of his death, always they marvel that one so close to bed, warmth, food, perhaps family, could stray so easily into danger.

My son, ten years old, is not worried at all. I make him stop to drink water and eat chocolate. I remind him to let me know at the first hint of a chill, of cold toes, of stiff fingers. I want to know right away, before he becomes really cold. Later, he complains of cold cheeks, and we stop. I check his face for patches of white, then rub my fingers across his skin looking for hard patches, for any sign of stiffness. There is nothing.

"You told me to tell you right away," he says. "But I'm not that cold."

We ski on.

❄   ❄   ❄

In 1911, Robert Falcon Scott was in Antarctica, waiting for summer so that he could start what would prove to be his fatal walk to the South Pole. Scott sent Apsley Cherry-Garrard, Edward "Bill" Wilson, and Birdie Bowers to Cape Crozier in July, the height of the Antarctic winter. Their mission: to bring back eggs from an emperor penguin colony. "They are extraordinarily like children," Cherry-Garrard later wrote, "these little people of the Antarctic world, either like children, or like old men, full of their own importance and late for dinner, in their black tail-coats and white shirt-fronts—and rather portly withal." The men hauled a sled sixty miles in the dark at temperatures of seventy below. Their tent blew away in a storm, and for a time the men lay in a snowdrift singing hymns and waiting to die. But in the end, they survived. They observed the birds at the nesting colony. Cherry-Garrard felt compelled to mention the tenacity these birds showed in caring for their eggs and young. "Now we found that these birds were so anxious to sit on something," he wrote, "that some of those which had no eggs were sitting on ice!

Several times Bill and Birdie picked up eggs to find them lumps of ice, rounded and about the right size, dirty and hard."

The men carried three eggs back to Scott's base, and the eggs now sit in London's Natural History Museum. Emperor penguins may look ridiculous, standing in the cold, mistakenly sitting on rounded blocks of ice, but Cherry-Garrard and his companions must have looked more ridiculous still, pulling their sled through the dark, resting in frozen sleeping bags, very quickly reaching a point at which they wished for the comfort of their own deaths. The penguins, meanwhile—without bags, without parkas, without kerosene or a stove to burn it in—maintained a body temperature of ninety-nine degrees. They did this despite the fact that emperor penguins go without feeding for as much as four months during the winter. They stand with their feet tipped up so that only their heels touch the cold ice. A heat-exchange system in their nostrils, a variation of the rete mirabile seen in the blubber of a whale, captures eighty percent of the heat that would otherwise be lost with each exhalation. As many as six thousand of them huddle through the winter in single groups, conserving heat, each bird leaning slightly forward, maintaining pressure on the one in front of it, creating in the end a circulating mass of penguins so that no one animal is stuck on the outside long enough to freeze. Eventually, under the care of the males, chicks hatch. A penguin chick is naked, unfeathered from the neck down, so it spends its first days standing on its father's feet, tucked into a brood pouch at the base of the father's rather portly belly. Although a sixty-mile hike almost killed Cherry-Garrard and his companions, and likely would have killed almost any other humans who had attempted it with the equipment available to these men, the penguins stand around incubating eggs and raising their young without feeding for months on end before marching fifty or sixty miles back to the sea, where they will end their four-month fast by plunging into ice-cold water and chasing down crustaceans and fish.

Nearly two hundred thousand emperor penguins live in Antarctica. They are, arguably, the most remarkable of the winter-active birds, but they are hardly the only winter-active birds. Redpolls, ravens, chickadees, ptarmigans, and a host of others prefer the risks of cold to the risks of migration. Of birds that overwinter in the cold, only the poorwill is known to enter into a state of long-term hibernation. The others have various tricks to survive the cold. The ptarmigan, best described as a snow chicken, a near cousin of the prairie grouse, has feathers that cover its feet and toes. In winter, when it has to digest fibrous twigs and bark, its gut lengthens. It hunkers down in the snow, only to flush out suddenly, flying low, landing not too far away, white feathers disappearing into white background. It goes beyond hunkering, knowing how to avoid wind by tunneling into snow, forming a cave near the top of a curving passage that traps its body heat. It hides out with the lemmings and voles and bugs beneath the snow, but unlike these subniveans, it uses its tunnels intermittently. It comes out to feed or, like the raven, to take a snow bath, throwing snow into the air and rubbing it across its body.

A naturalist writing in 1900 described insulation found in a crossbill's nest, saying that the nest was lined with "long black tendrils resembling horse hair." Another, writing nine years later, described the insulation as "wool and moss" and "rabbit fur."

Other birds shiver through the winter. Shivering, though, requires calories. A chickadee's feeding rate increases twentyfold in winter. A crossbill needs to find a spruce seed every seven seconds. An emperor penguin might start the winter at eighty pounds and end it at a lean fifty, emaciated under its feathers.

It is the feathers that allow all of these birds to remain active through the winter. Without feathers, none of them would have a chance. Pennaceous feathers — feathers that can become quill pens — cover these birds. Feathery barbs extend from the central shaft of pennaceous feathers, and along these barbs are microscopic barbules and tiny hooks called barbicels. The barbicels lock one feather to the next, encasing it, superior to shingles covering a house. Dry beneath the pennaceous feathers, down feathers — fluffy feathers with no barbules, the kind of soft, tiny feathers that are used to stuff jackets and expensive sleeping bags — hold air and warmth. Even when the bird dives, the down, beneath the interlocking pennaceous feathers, remains dry.

If the pennaceous feathers leak, the down becomes wet and useless. A bird with leaking pennaceous feathers cools quickly. In oil spills, thousands of birds have found themselves swimming in crude. The oil makes their pennaceous feathers leak. They try for a time to maintain body heat by shivering, first in their pectoral muscles, where the wings attach, and then later in their leg muscles. Shivering burns stored calories. While they shiver, they preen, trying to scrape away the oil. When the stored calories are gone and they can shiver no more, they die of hypothermia. Or is it starvation?

It has been said that feathers evolved first to protect birds from the cold, and later the birds realized or learned or somehow found out that feathers, managed in just the right way, would allow the convenience of flight.

JANUARY

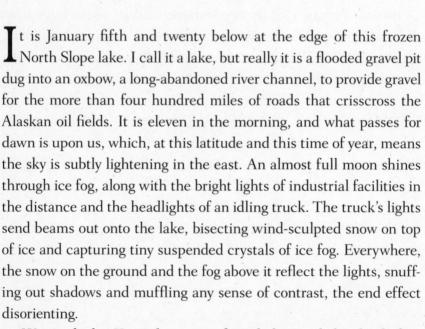

It is January fifth and twenty below at the edge of this frozen North Slope lake. I call it a lake, but really it is a flooded gravel pit dug into an oxbow, a long-abandoned river channel, to provide gravel for the more than four hundred miles of roads that crisscross the Alaskan oil fields. It is eleven in the morning, and what passes for dawn is upon us, which, at this latitude and this time of year, means the sky is subtly lightening in the east. An almost full moon shines through ice fog, along with the bright lights of industrial facilities in the distance and the headlights of an idling truck. The truck's lights send beams out onto the lake, bisecting wind-sculpted snow on top of ice and capturing tiny suspended crystals of ice fog. Everywhere, the snow on the ground and the fog above it reflect the lights, snuffing out shadows and muffling any sense of contrast, the end effect disorienting.

We are lucky. Yesterday it was forty below and the day before even colder, the temperature as brutal as a physical assault, making one gasp for air but then forcing a stop mid-gasp as cold air batters warm lungs. The difference between forty below and twenty below is striking. Today, at twenty below zero, it is cold enough only to ice up one's eyelids and freeze the inside of one's nose.

We walk out onto the lake. The snow screeches under our boots, as if we are walking on Styrofoam while wearing Styrofoam boots. It is so cold and dry that our boots kick up little clouds of snow that the breeze carries along the surface of the lake, like dust clouds one might kick up on a desert sand flat. Fox tracks cover the lake, zigzagging from one side to the other, looping around on themselves. There are rabid foxes in the area. Just a week ago, a fox ran repeatedly out onto a gravel road, attacking anything that moved, including trucks. When the fox attacked a front-end loader, it lost, and its crushed carcass, sent to a state laboratory, tested positive for rabies.

I am with a team of hydrologists. They want to understand how water chemistry changes through the year, and especially in winter, when lakes are hidden by snow-covered ice. They want me to understand what they are doing and the difficulties they face. They drag two sleds onto the lake. A woman carries shovels and a gasoline-powered drill, and another woman holds a tent of the kind used by whiskey-sipping ice fishermen in Minnesota. The hydrologists shovel snow from the lake's surface, pull-start the drill motor, and bore down into the ice. Though the snow dampens it, the noise of the two-cycle engine fills the cold air. In less than a minute, the drill cuts in a foot, two feet, three feet, and then water gushes up and spills out across the surface of the ice, turning immediately to slush and within seconds forming clear ripples of new ice across the older translucent ice of the lake's surface. The hydrologists erect the tent over the hole. Inside, they light a small heater. The heater's immediate impact on the warmth of the tent is comparable to what one would expect from a birthday candle. It is so cold in the tent that slush forms continuously in the hole. They scoop the slush from the top of the hole and lower an instrument thirty feet into the water.

They measure temperature, pH, conductivity, and dissolved oxygen. The heater glows in the corner of the tent, but it is pathetically underpowered and effectively heatless, providing only a soft red glow and offering nothing more than ambiance as one hydrologist reads off data and the other sits on a box taking notes.

It is oxygen that interests the team most. There are fish in this lake, grayling that swam in from the adjacent river during high water. They overwinter in these deep lakes, these flooded gravel pits, where they find water that will not freeze solid by spring. This lake will not freeze solid. In all likelihood, the ice will grow six feet thick—somewhat thicker than most hydrologists are tall. Below the ice is fifty feet of liquid habitat. But the ice seals the surface. Whatever oxygen was in this lake when it froze is all the oxygen these fish will share until spring. At the bottom of the lake, oxygen-hungry bacteria work the sediments, decomposing anything that has fallen down from above. The bacteria compete with the fish for oxygen, sucking it out of the depths.

"These lakes," the hydrologist tells me, "provide water for industry." The water in these gravel pit lakes is used for drilling and for washing trucks. It is used for building temporary roads of ice that reach out across the tundra to winter construction projects. It is even used for drinking and bathing. The fish are kind enough to give up millions of gallons each year, to share it with the oil industry. The water level of this lake will drop five feet before spring, those five feet hauled away ten thousand gallons at a time by big red tanker trucks. Even now, as we talk, a truck sucks water from the lake. The driver sits in the cab, bored but warm, with a black hose running to a white pump house next to the lake and steamy exhaust from the truck blowing sideways, out over the lake, toward our little tent with its glowing, insignificant heater and its scientists armed with instruments and data sheets and thick gloves. Under all of this ice, the fish tread water, waiting for spring, their chilled brains probably incapable of even wondering what that funny pumping sound is that they hear now and again as these humans come and suck out five feet of swimming headroom.

Later, in the warmth of a dining hall, a man tells me that a worker has been attacked by a fox. It happened ten miles from our gravel pit lake. The fox ran out of the darkness and bit the man in the leg. The man tried to shake it off, as one might shake off a pit viper, but the fox hung on to the cuff of his snow pants. He kicked it. It lay in the snow as if dead but later got up and staggered off into the Arctic winter. Its rabid teeth, slowed down by the worker's snow pants and stopped by his boots, never penetrated the man's skin.

❄   ❄   ❄

The world's liquid freshwater resides in rivers and lakes, in underground aquifers, and in the sky. Eight-tenths of the world's freshwater is frozen. Another three thousand trillion gallons drifts in the atmosphere. But to say that it drifts does not do it justice. The water cycles through, riding thermals, now as a vapor, now as a droplet, now as a speck of ice coalesced around a particle of dust nine miles high in the troposphere. It is now in the air as mist, now as rain within a cloud, now in the Pacific, now as a downpour falling on hot pavement in south Georgia, now on the northern pack ice. It might rise up from the equator, sink down as a driving rain in the horse latitudes, drift for a while in the Gulf Stream. Rising again to travel farther north, it eventually makes its way poleward and falls as snow, then melts, running off into the North Atlantic to follow currents carrying it downward, finding its way to an upwelling, then drifting on the surface until it evaporates once more, only to wind up God knows where. In geological time, water molecules have been grand travelers, each finding its way everywhere, touching down everyplace, like irrepressible tourists on a four-and-a-half-billion-year junket.

To understand the movement of water, to understand weather, one has to appreciate the earth for what it is: a spinning round ball with a rough surface. If the world were flat, facing the sun, receiving equal amounts of sunlight across every square inch, it would be a simpler place. It is not flat. It is a sphere. A square foot of sunlight hitting the equator at noon spreads out across a square foot of the earth's surface, while a square foot of sunlight hitting the earth near the poles, where the globe curves away and the earth's surface is turned at an angle to the sun, spreads out over two square feet or three square feet or ten square feet of the earth's surface, depending on where one stands. The square foot of light and heat and sustenance that hits the ground at the equator has to be shared across those multiple square feet at higher latitudes.

Air at the equator, warmed by the sun, rises. Rising air leaves behind an area of low pressure — a low, as it is called. Wind blows from areas of high pressure to areas of low pressure, performing the singular role of restoring equilibrium, of preventing too much air from piling up in any one place. But in fulfilling this role, wind transfers heat. Near the surface of the earth, air rushes in to fill the low, replacing the warm air that has risen. The new air itself warms under the tropical sun and rises. More air is sucked in. Meanwhile, the rising hot air spreads out as it gains altitude. In spreading out, its pressure drops, and the heat contained in the air mass spreads out too, making the air mass cooler. The drop in pressure is accompanied by a drop in temperature. The whole mass spills outward from the equator. Along the way, water vapor carried in the air mass grows cool enough to condense and tumble downward as liquid water. Around the latitude of Shanghai and Jacksonville in the Northern Hemisphere and Easter Island and Cape Town in the Southern Hemisphere, and moderated by local geography and the myriad factors that affect air movements, it tumbles down. George Hadley imagined these global patterns in 1735, before satellites, before computers, before reasonable maps of the world. The global loops of rising and cooling air near the equator became known as

Hadley cells. Farther north, similar patterns of rising and falling air became known as Ferrel cells and Polar cells.

The earth, spinning, moves beneath the air above it. The air—cycling up and down in Hadley cells and Ferrel cells and Polar cells, then spilling out north and south in what should be a straight line—is turned by a spinning earth. In the Northern Hemisphere, the earth's spin tends to move wind to the right of its direction of travel. In the Southern Hemisphere, the earth's spin moves wind to the left of its direction of travel. This effect, this odd rightward and leftward trending of moving air, was described in 1835 by the Frenchman Gaspard-Gustave de Coriolis and has become known as the Coriolis effect.

But the earth is rough, with mountains and valleys and their attendant shadows. Here on the southern slopes of this mountain, the sun bathes the earth in warmth, but there in that shadowed valley, the earth is cool. And the ground itself is patchy. Here on this dull patch of bare dirt, sunlight warms the soil, while there on that patch of snow—on that patch of crystalline water turned white and smooth to form what amounts to the closest thing nature offers to a perfect reflector—the warmth bounces off, back into the sky. Under this clear blue sky, the heat is lost, reflected back into space. There under that cloud, the heat is trapped, held in by a blanket of dust and moisture. This shoreline warms quickly under the morning sun, sending its air skyward, and the air above the ocean or lake or river blows shoreward to fill what would otherwise become a vacuum. The air above that black roof is hot, and when it moves skyward, it sucks in air from around the yard, which then is heated and sent skyward, too. The air is heating and cooling and tumbling about, cells within cells within cells, none of it standing still for very long, all of it moving with a Coriolis twist.

Even within the simplicity of Hadley cells and Ferrel cells and Polar cells, ignoring the spinning earth and the irregularities of mountains and reflections from snow, local complexities arise. Superimposed on the simplicity of global patterns is the nature of fluid dynamics. High in the atmosphere, where warm and cold air meet, vortices form, like the eddies and whirlpools of fast-flowing rivers, spinning around themselves and floating downstream, confusing the eye by combining directional motion with spinning and chaotic dancing. The eddies become regions of low pressure, depressions that must be filled. They suck in air and moisture, pulling it skyward, and high in the sky condensing damp air to rain or snow or sleet or hail and then tossing it back to the earth.

Wind moves frigid air to warmer climes. It creates blizzards that trap schoolchildren on the prairie. It creates raging gales into which people walk or sail or ski. It picks up snow that sand-blasts bark from trees.

In the end, weather can be described as a mishmash of events, each one alone predictable, but intermingling to compound one another and confuse the issue, and in the end adding up to nothing less than a complex mess of unpredictability.

The ancient Babylonians said, "When a halo surrounds the sun, rain will fall. When a cloud grows dark in the sky, the wind will blow." Before Socrates, Thales of Miletus made a weather calendar. Aristotle commented on clouds, dew, snow, and hail, recognizing that they differ because of temperature. The barometer was invented in 1643 and the anemometer, for measuring wind speed, in 1667. Ben Franklin realized that the weather in Philadelphia came from somewhere else and left for somewhere else. His attempts to observe a lunar eclipse in 1743 were foiled by storm clouds, but his friends in Boston watched the eclipse and then, four hours later, watched his storm clouds roll in.

By 1846, weather reports transmitted by telegraph could be purchased for between twelve and twenty-five cents a day. During the Crimean War, the warship *Henri IV* was lost in a storm on November

14, 1854, and Urbain Leverrier, director of the Paris Observatory, urged the French government to recognize the need for improved weather forecasting. A year later, in the United States, the Smithsonian was posting weather maps in its Great Hall. Networks of weather reporters—some paid, some amateurs—sent information on local conditions to central repositories. A man wearing a raincoat and carrying an umbrella might sit on a park bench in the city, studying the contents of a rain gauge, while another might record temperatures on Texas rangelands from the back of his horse. A third might measure the wind blowing in off a busy harbor, and a fourth might record the presence of morning dew on his cornfield. And then, with all of these observers working, with all of them piping in information through more than twenty thousand miles of telegraph wires, Adolphus Greely, not long back from the Arctic, failed to effectively foresee the Blizzard of 1888, the School Children's Blizzard, predicting instead a cold wave with snowdrifts. The failure left nineteen-year-old Etta Shattuck alone for three days, bivouacked in a haystack, singing hymns and praying while the storm raged, saved from the haystack only to die from the infections that followed frostbite. The failure left a seventeen-year-old girl frozen to death standing up. Because of the failure, the bodies of the Kaufmann brothers, who died huddled like penguins trying to stay warm, had to be thawed in front of a woodstove before they could be separated.

Vilhelm Bjerknes, a Norwegian working at the beginning of the twentieth century, was the first to propose the application of thermodynamics and fluid mechanics to the atmosphere. His thoughts evolved to rely on a system of cells, stacked one above the other and covering the entire earth in nothing less than a three-dimensional checkerboard. The idea was to populate the three-dimensional checkerboard with data from observations and then use the data to predict what would happen next.

During World War I, the English meteorologist Lewis Fry

Richardson tackled the rat's nest of calculations needed for numerical forecasting. In 1922, he published a book saying that the calculations would require sixty-four thousand people working day and night to keep up with the weather. He envisioned a city of workers in a building laid out to mimic the globe itself, with each of the workers struggling through his equations in a space representing his part of the globe. There would be green space outside, soccer fields and lakes. Those who predicted the weather, Richardson believed, should have the opportunity to experience it. In the end, though, the city was never built. It turns out that this decision was justified. Had the city been built, it would have failed in its purpose, doomed from the outset by a naive belief in a strictly deterministic universe.

The Americans were the first to use electronic computers in weather prediction, in the 1950s. The data, one might think, would be adequate: more than ten thousand weather stations check conditions around the globe, another five thousand ships and planes send in information, unmanned buoys transmit data from remote reaches of the world's oceans, more than a thousand weather balloons go up each day to sample the sky, and satellites circle endlessly with their gaze turned back toward earth. But the data are not adequate. In 1963, Edward Lorenz set up weather models on a computer. He compared models run with data offering three decimal points of accuracy and those run with data offering six decimal points of accuracy. The results were completely different. Tiny differences in the starting point resulted in major differences at the end point. It would be comparable to a banker counting his wealth in dollars and in pennies, only to discover that he was well positioned in dollars but flat broke in pennies. It made no sense. It led to what was later called chaos theory. Lorenz delivered a talk to the American Academy for the Advancement of Science titled "Predictability: Does the Flap of a Butterfly's Wings in Brazil Set Off a Tornado in Texas?" The answer: yes. Or at least it might.

In medieval times, weather predicting was an occult art. Fore-casters were burned at the stake.

❄   ❄   ❄

It is January ninth and twenty below at the Anchorage airport, close to the record cold set in 1952. The ground hides under four feet of snow, with plowed piles and drifts running deeper. Long icicles hang from roofs. El Niño, where have you gone?

I head south. By the time I fly over the Canadian border, tem-peratures on the ground are above forty. They hover in the forties for more than a thousand air miles, and then, as abruptly as a color change on a weather map, they reach the fifties. And by the time I land in New Orleans, the mercury flirts with seventy. I have flown across almost ninety degrees of temperature change.

For the most part, the flight path, at thirty thousand feet, took me through the troposphere. Puffy white cumulus clouds call the lower troposphere home. Cumulonimbus clouds can start down at the level of cumulus clouds, but they tower skyward as much as six miles, with updrafts that send pellets of water screaming toward space, turning to ice or snow as the air cools, then plummeting back down—not drifting down with gravity, but flushing down in rushed gusts. They roller-coaster up and down, sometimes freezing and thawing repeatedly, maybe eventually breaking loose to parachute to the ground as rain or snow or ice or undecided sleet. A cumulonim-bus cloud can hold five hundred thousand tons of water.

For every mile upward in the troposphere, for every mile farther from the earth's surface, the temperature drops seventeen degrees, plummeting to 65 below. But then it rises again in the stratosphere, warming up to the freezing point in the blanket of ozone that drifts around between nine and twenty-five miles up. Beyond, it cools back down. Around the fifty-mile mark, near what most would consider the edge of space, the thermometer drops to 180 below. At this temperature, carbon dioxide freezes solid. A few miles farther up, where the northern lights dance but still well below the realm of weather satellites and space shuttle orbits, the temperature rises. It exceeds 1,000 degrees, hot enough to melt lead and zinc, but in air so thin that it does not matter, in air so thin that it is not worthy of the name.

A hundred and fifty years ago, in England, not far from Westminster Abbey and Windsor Palace, a man named James Glaisher amused himself by sketching snowflakes. Not satisfied with what he found on the ground, he strapped a basket to the bottom of a balloon, loaded the basket with a drawing pad and an assistant, and rode it upward. In warmer clouds, reasonably close to the earth and ripe with humidity, Glaisher sketched the star-shaped flakes of Christmas cards. As he went higher, the air cooled. His assistant grew cold. The assistant's hands, in particular, were chilled. Glaisher pressed on. The assistant, his loyalty guaranteed by the absence of any reasonable alternative to staying in the basket, stood by his side. Glaisher sketched hexagonal crystals of snow at five degrees above zero and column-shaped flakes at fifteen below. At twenty-nine thousand feet, in very thin air, Glaisher collapsed. The basket swung wildly beneath the balloon. Glaisher's assistant tried to release gas from the balloon, but his freezing hands were too stiff to pull the dump cord. He gripped it in his teeth and pulled. Gas flowed out of the balloon, and both men survived.

Glaisher wrote of the snowflakes he had seen: "Their forms are so varied that it seemed scarcely possible for continuous observations to exhaust them all."

I amuse myself on the airplane with a collection of quotations about weather lore:

A bad winter is betide,
If hair grows thick on a bear's hide.

If onions are more abundant than bears, there is this:

Onion skins very thin,
Midwinter coming in;
Onion skins very tough,
Winter's coming, cold and rough.

Or this gem about February second, Candlemas Day, still three weeks off:

If Candlemas Day be fair and bright
Winter will have another fight.
If Candlemas Day brings cloud and rain,
Winter won't come again.

❄ ❄ ❄

In 1776, a son of the parish clerk of Bampton in Devon, England, was killed by an icicle that plummeted from the church tower and speared him. His memorial:

Bless my eyes
Here he lies
In a sad pickle
Kill'd by an icicle

On August 16, 1970, a chunk of ice fell from an airplane and crashed through the roof of a home just outside London. On March 25, 1974, ice eighteen inches across slammed into the hood of a woman's car, again near London. She was later compensated by an airline. In March 1978, Chicago police sealed off roads around the city's tallest buildings while ice, accumulated during a storm, crashed to the sidewalks.

On March 7, 1976, in Virginia, a basketball-size chunk of ice crashed into a roof, but this time there were neither airplanes nor skyscrapers anywhere in the vicinity. On June 4, 1953, in southern California, fifty lumps of ice fell, weighing in total about a ton and with individual pieces as heavy as an adult man. Farther back, on August 13, 1849, a block of ice nearly seven feet in diameter fell in Scotland. According to an 1849 issue of the *Edinburgh New Philosophical Journal*,

> a curious phenomenon occurred at the farm of Balvullich, on the estate of Ord, occupied by Mr. Moffat, on the evening of Monday last. Immediately after one of the loudest peals of thunder heard there, a large and irregular-shaped mass of ice, reckoned to be nearly 20 feet in circumference, and of a proportionate thickness, fell near the farm-house. It had a beautiful crystalline appearance, being nearly all quite transparent, if we except a small portion of it which consisted of hailstones of uncommon size, fixed together. It was principally composed of small, square, diamond-shaped stones, of from 1 to 3 inches in size, all firmly congealed together. The weight of this large piece of ice could not be ascertained; but it is a most fortunate circumstance, that it did not fall on Mr. Moffat's house, or it would have crushed it, and undoubtedly have caused the death of some of the inmates. No appearance whatever of either hail or snow was discernible in the surrounding district.

The May 1894 *Monthly Weather Review* reported an ice-encased gopher turtle falling during a hailstorm in Bovina, Mississippi, and in December 1973, a newspaper reported frozen ducks falling in Stuttgart, Arkansas.

And then there is snow. The journal *Nature* reported three-and-a-half-inch flakes from a 1997 storm. In January 1915, snowflakes three and four inches across fell on Berlin. According to the *Monthly Weather Review* of February 1915, the flakes "resembled a round or oval dish with its edges bent upward." And on January 28, 1887, a report from Montana described flakes — "flakes" in this case perhaps an odd choice of word — fifteen inches across and eight inches thick.

The chaos of weather spills over with freakish events. But it is usually the merely unusual ones, not the freakish, that make history. There is, for example, nothing freakish about hail. It forms regularly in cumulonimbus clouds, with little balls of water and ice riding winds skyward, reaching altitudes beyond the realm of jet planes and temperatures of one hundred degrees below zero, often falling and rising many times, buffeted by the internal chaotic gales of cumulonimbus thunderheads, but finally falling from the sky.

There are records of freakishly big hailstones: A 1697 hailstorm in England dropped four-inch hailstones that killed at least one person. A hailstone in Kansas weighed just under two pounds. The largest recorded hailstone, weighing more than two pounds, fell in 1896 in Bangladesh. But it is the lesser hailstorms that leave historical footnotes: A hailstorm in April 1888 killed 246 people in India. In April 1977, a hailstorm took out the engine of an airplane and smashed its cockpit window, killing 68 people after a crash landing on a Georgia highway. And in 1984, a hailstorm caused well over a billion dollars' worth of damage in Munich.

As for snow, the School Children's Blizzard of 1888 was somewhat unusual with its sudden brutality, but it was hardly freakish. Two years earlier, a blizzard hit the western Texas Panhandle, Indian territory that became Oklahoma, and Kansas. Afterward, dead cattle littered the land. In 1887, another blizzard hit ranching country, this time in the Dakota Badlands. Thereafter, ranching changed forever, featuring smaller herds of higher-quality animals that were more tightly controlled. And just after the School Children's Blizzard, also in 1888, a low-pressure system moved into New York City from the Atlantic, dumping snow and pushing winds to seventy miles per hour. Greely's team of weather forecasters missed the call again, predicting rain and "colder fresh to brisk westerly winds, fair weather." The storm surprised the city, and more than two hundred people died. From the *New York Evening Sun*:

> The streets were littered with blown down signs, tops of fancy lamps, and all the wreck and debris of projections, ornaments, and moveables. Everywhere horse cars were lying on their sides, entrenched in deep snow, lying across the tracks, jammed together and in every conceivable position. The city's surface was like a wreck-strewn battlefield.

From the *New York Tribune*: "The city was left to run itself, chaos reigned, and the proud boastful metropolis was reduced to the condition of a primitive settlement." And from the *New York Times*:

> In looking back on the events of yesterday, the most amazing thing to the residents of this great city must be the ease with which the elements were able to overcome the boasted triumph of civilization, particularly in those respects which philosophers and statesmen have contended permanently marked our civilization and distinguished it from the civilization of the old world—our superior means of intercommunication.

Ice falls on people and airplanes, snowstorms seize cities, and cold snaps win and lose wars. Snow stopped Alexander the Great's eastward march into India three centuries before the birth of Christ, and it blocked the Moors' invasion of France in the thirteenth century. It added to the suffering of George Washington's twelve thousand ill-prepared Continental soldiers at Valley Forge, prompting Gouverneur Morris of New York to describe the men as "an army of skeletons." In yet another conflict, Napoleon's soldiers reached Moscow in mid-September 1812, a year rendered somewhat colder than normal by an atmosphere laced with volcanic dust. By early November, temperatures were below zero, and by early December the French were retreating at thirty-five degrees below zero. Still later, during World War I, Italian and Austrian soldiers used avalanches as weapons, killing an estimated sixty thousand enemy troops. Bodies were still turning up as late as 1952.

Hitler's 1941 invasion of Russia faced snow in October. German land mines failed because of snow and ice. Russian artillery troops used lubricants suited for low temperatures, while German soldiers had to warm their artillery with campfires. The Russians used ponies acclimated to winter, and many of the Russian soldiers knew how to ski. German tanks bogged down in the snow. At temperatures of forty-nine degrees below zero, the German soldiers awaited winter clothes. It is said that a quarter of a million German soldiers died of frostbite and hypothermia. Cold was an ally of the Russians.

Superimposed on cold weather and its freak events, on all of the difficulties of prediction and the dreams of solving unsolvable equations and on the beating of a butterfly's chaotic wings, discernible patterns remain. Air moves irrevocably from areas of high pressure to areas of low pressure. It is easy enough to predict the weather a day or two out by plotting the motion of fronts and knowing, more or less, how one will interact with another. In more general terms, there are Hadley cells and trade winds. There is the Coriolis effect shifting air to the right and left as wind moves across the rotating earth. There is El Niño. In the north, there is the Pacific Decadal Oscillation, shifting phases every twenty years or so. During its positive phase, the western Pacific becomes cool, and part of the eastern ocean warms. During its negative phase, the western Pacific warms, and the eastern ocean cools. Farther north, there is the Arctic Oscillation. In its negative phase, counterclockwise winds blowing in the stratosphere weaken, and high pressure stands over the

Arctic, pushing frigid winter air farther south, generating rain in the western United States and the Mediterranean, and weakening trade winds. In its positive phase—the phase in which we have been stuck more on than off for the past twenty years—the stratospheric winds blowing counterclockwise above the pole strengthen, middle America stays warmer, and California and Spain dry out.

And there is the most basic pattern of all: polar regions are cold, and tropical regions are hot. The sun is spread out across the polar regions, its light and heat striking at an angle. Most of the energy that reaches a polar surface bounces back, reflected by snow and ice back up into space. In the tropics, the light and heat hit head-on. The ground absorbs the heat. Leaves absorb the heat. Water absorbs the heat. And then the polar and equatorial regions interact. On a global scale, seen from a distance, it might be said that the polar regions suck in the heat of the tropics, swallowing the world's warmth. The equatorial regions shed heat south and north, like a Weddell seal steaming as it lies on the Antarctic ice, or like a moose panting, overheated and uncomfortable, its hot breath projecting vaporous shadows against the snow.

❄ ❄ ❄

It is January twenty-fifth in Barrow, Alaska, the northernmost community in the United States. The forecasters' prediction of a few days of warm weather has proved true. At fifteen below, it is a mild January afternoon. Two days ago, the sun rose for the first time this year, but today is the first day with clear skies, and the first visible sunrise since mid-November. The sun eases upward, then hovers, moving in a shallow, graceful arc from east to west, never more than one or two degrees over the horizon. We are on snowmobiles, and we ride for some time into the sun. I leave a hundred feet between me and my companion's profile, his hunched form on the machine's saddle. I see his parka hood framed against the Arctic sun, the silhouette of a shotgun strapped to his back, and a light cloud of condensation and fumes from his exhaust pipe, all riding ahead of his long, stretched-out shadow on the snow. Around him, the flat white snow and ice, devoid of contrast, confuses the eye. I scan occasionally for arctic foxes, but see none. There are fewer foxes here, in Barrow, than in the oil fields southeast of here. Although we travel armed, it is more from habit than from need. The time to see polar bears is spring, summer, and autumn, not January. Now they are scattered on the sea ice or denned up to give birth and suckle their young, and in any case their reality is less ferocious than their reputation.

The tundra in panoramic view appears flat, but on the machines we feel every bump. On occasion, we dip down a few feet or ride over a shallow hill. With each dip we see the sun set, and with each hill we see the sun rise. It is an orange orb, angling in through low haze on the horizon, the sky above it open and deep blue. We are riding into the wind. The cold nibbles and then chews at exposed flesh around my cheeks and temples. At twenty-five miles per hour, the windchill equivalent is minus forty-four. Any warmth from the rising sun is more psychological than real.

We intercept an ice road and ride along its smooth surface, picking up speed and occasionally fishtailing on the ice. Ice roads, built by dumping water onto the snow covering the tundra, or right on the tundra itself, then scarifying the ice for traction, are used throughout the far north for winter travel. This particular ice road heads far off into the tundra to an exploration well. Someone hopes to find natural gas there. It will provide a backup supply for Barrow, a gas-rich community. A water truck drives by, and we have to dodge momentarily into the tundra, then come back onto the road. The truck looks ludicrously small against the ice road itself, and the ice road, disappearing toward the rising sun, looks ludicrously small against the expanse of snow-covered tundra.

We cut off from the road, ride across a frozen lake, and intercept the coast. Although my face stings with cold, my body is overheating, like that of a running moose or an agitated bowhead whale. I unzip the top few inches of my parka. The early exhilaration of the ride has given way to irritation with the bumpiness and the cold wind and the restricted vision of goggles and a parka hood. We stop at the coast and turn off our machines. Immediately, the exhilaration returns. My companion has ice on his beard and collar. To our right, we can see Barrow, its five thousand residents scattered in village sprawl, picturesque against the frozen sea. Closer, between here and Barrow, someone has stored a boat on the tundra. It is upside down, standing on four steel fuel drums, frozen in place and masked with frost. It is a skin boat, an *umiaq,* built from the stitched hides of bearded seals and used to hunt bowhead whales in the open-water leads during spring, when the whales are swimming east between ice floes and snorkeling from pool to pool. To our left, we see the coastline, empty but for ice rubble, bulldozed into piles by the slow but powerful movements of a frozen ocean. The Inupiat call the piled ice rubble *ivuniq.* This rubble, standing no more than fifteen feet tall, might be called an *ivunibauraq,* a little ice ridge. *Ivuniq* forms when sheets of sea ice, miles across, pushed by wind and current, slowly collide. The leading edge crumbles against the shore or against another sheet of ice, piling up into a ridge, like drifting tectonic plates forming frozen models of the Himalayas and the Andes and the Rockies in fast-forward. At times, the thunder of ice collisions can be heard for miles through the still, cold air.

Barrow, like most communities in Alaska, looks temporary, like a pioneer settlement. It is not. Barrow is among the oldest permanent settlements in the United States. Hundreds of years before the European Arctic explorers showed up, starving and freezing and succumbing to hardship, Barrow was more or less where it is now, a natural hunting place at the base of a peninsula that pokes out into the Beaufort Sea. It was called Ukpeagvik, literally "a place for hunting snowy owls." Yankee whalers sailed here, learning about the bowhead whale from Inupiat hunters, but also bringing new harpoons, steel, and guns. A weather station of sorts was established in 1881. Later, the military came, setting up a radar station, and in 1947 a science center was founded at Barrow. Men raised as subsistence hunters showed scientists how to function in the Arctic. They shared traditional knowledge. They corrected the misconceptions of what has come to be known regionally as "Western science." Today Barrow has the coldest average yearly temperature of any community in the United States, at just under ten degrees.

We look out over the frozen ocean and see nothing but wind-sculpted waves of snow and ice. Though invisible, there are also seals and bears and arctic foxes, and farther out, the North Pole. Soft light comes in low and angles across the ice. We stare at the northern ice cap, a reflector the size of a continent that bounces what little sunlight it receives back into space, an ice cube proportioned to cool the entire globe. There is nothing more to see than a rough white surface disappearing into the horizon, yet we stand silently for some time, concerned that in turning away we might miss something very important, something crucial to our well-being and somehow central to our lives.

FEBRUARY

It is February second, Candlemas Day, and a sweltering forty-eight degrees here in Anchorage. It is eleven degrees in Kansas City, thirty-seven degrees in Washington, D.C., and twenty-six degrees in Denver. New Orleans hit only forty-nine, and the low in Los Angeles was colder than the high in Anchorage. My beautiful snow is melting again, filling the streets with slush and water. Across town, a creek thawed and flooded the basement of an office building, ruining computers. Roofers are busy fixing leaks of suddenly liquid water. At the Alaska Zoo in Anchorage, Jake the brown bear woke up, groggy, and staggered outside to lie in the sun.

As the snow melts, it grows gritty with a few months' accrual of dirt and dust, previously scattered through three feet of snowpack but now accumulating right on the surface. With a darker surface, the snow absorbs heat and melts that much faster. I ski in a T-shirt, moving along a trail on the edge of Cook Inlet, watching thousands of blocks of dirty ice float past with the tide. Hot air blows in from the south, stripping away the cold or, closer to the truth, pumping in heat. Right on the edge of the inlet, slushy puddles that tug at my skis are interspersed with the sort of crusty snow that comes from freezing and warming and freezing again. My dog, running along behind me, breaks through the crust and looks at me quizzically, head tilted and ears up, seeming to wonder when we might turn around. But under trees, in the protection of windbreaks, the snow remains firm.

I take heart. "It's definitely a warm event," Sam Albanese of the National Weather Service tells reporters, "but it's certainly not out of the realm of what happens most winters here." In February, in Anchorage, one can be certain that the cold will return. Mary Anderson, who has lived in the area since 1945, tells a reporter, "I've seen every kind of weather you can imagine. This isn't unusual."

It is in fact somewhat unusual. It is a record high for this date, another record high in the annals of global warming. The previous high was forty-six degrees in 1977. On this same date in 1947, the city enjoyed a record low. It was thirty-three degrees below zero.

Candlemas Day marks the halfway point between the winter solstice and the spring equinox — the halfway point between the shortest day of the year and the day with twelve hours between sunrise and sunset. It was once considered the last day of Christmas. Although now thought of as a Catholic holiday, it may have its roots in the Gaelic festival of Imbolc, celebrated long before Christ to mark the birth of spring lambs and the first stirrings of spring. *The Catholic Encyclopedia,* first published in 1907, denies this possibility so wholeheartedly that one is left convinced of its truth. The name Candlemas Day refers to the blessing of the candles to be used at Mass, but the day is also marked by the burning of candles in windows. The day has other names, including the Presentation of Christ in the Temple and the Purification of the Blessed Virgin Mary. And Groundhog Day.

In Anchorage, the sky is clear one minute and overcast the next.

The groundhog, known in Alaska as the marmot, might or might not see its shadow. One might or might not think of the day as clear and bright. Certainly, the shadows are not thick, and there is no rain. If the proverbs hold true, winter will be around for a while longer. Marmots, if any have broken hibernation, will waddle back into their holes, curl up, and drift back into their winter stupor. Their chubby little bodies will drop to within nine or ten degrees of freezing.

Coming down a small hill, I round a corner and am suddenly exposed to the wind from Cook Inlet. Icy snow turns to slush under my feet, ripping my skis out from under me. I go down on my hands and knees in the slush, my skis akimbo. Annoyed, I curse El Niño and chinook winds and Hadley cells. I curse global warming. I long for the cold of 1947.

❄   ❄   ❄

Young Frederic Tudor had a penchant for losing money. In February 1806, in his midtwenties, he filled a sailing vessel with ice from a Massachusetts pond and sent it to the island of Martinique. The *Boston Gazette* covered the story: "Loading ice — No joke. A vessel with a cargo of 80 tons of ice has cleared out from this port for Martinique. We hope this will not prove to be a slippery speculation." It did in fact prove slippery. Island people had no experience with ice. In a letter to his brother-in-law, Tudor wrote of customers who left their ice in the sun or in a tub of water and then complained when the ice melted. By 1812, Tudor was in debtors prison.

Three years later, out of jail and with more borrowed money, Tudor invested in an icehouse in Cuba. Pursued by sheriffs because of his debts, he set sail from New England on November 1, 1815. He spent the next ten years building the trade and learning how to preserve ice. He built icehouses throughout the Caribbean. He experimented with rice chaff, wood shavings, and sawdust as insulation. He created a demand for cold drinks in the tropics, which in turn created a demand for more and cheaper ice, which in turn inspired one of his suppliers to harness a horse to an ice saw, creating what would be called the ice plow and tripling production. By 1833, he was shipping ice to India, sixteen thousand miles and four months away from Massachusetts. This was at a time when many Indians had never seen ice. By 1855, Tudor would make more than two hundred thousand dollars from the Calcutta ice trade. He became known as "the Ice King."

None of this happened in a vacuum. The Inuit of the far north had been using ice cellars for thousands of years to preserve meat through the short Arctic summer. The Chinese had cut and stored winter ice since at least 1000 B.C. By about 500 B.C., the Egyptians were making ice in earthenware pots left out on cold nights. Ice, harvested in winter and stored in ice cellars or pits, had been used since at least Roman times to cool wine. In 755, Khalif Madhi used snow to refrigerate items that he carried across the desert. Giambattista della Porta made ice sculptures and served iced drinks in sixteenth-century Florence. By Tudor's time, every temperate zone town or village in the United States had at least one icehouse to store frozen pond water through the summer. In 1803, three years before Tudor sailed for Martinique with his load of ice, a man named Thomas Moore patented an icebox, a tight sheet-metal affair surrounding a cedar tub lined with rabbit fur and filled with pond ice. Moore called his icebox a refrigerator. He used it to transport butter to Washington, D.C. Where others were selling soggy gobs of butter in the swampy heat, he could sell attractive chunks of solid chilled butter for top dollar. He wrote *An Essay on the Most Eligible Construction of Ice-Houses; also, A Description of the Newly Invented Machine Called the Refrigerator.*

Three decades later, another inventor, Jacob Perkins, made the first practical device that we would think of as a refrigerator today. Different models were developed and marketed, but they all used the same principle. A liquid running through tubes was allowed to vaporize inside of an insulated box, and in vaporizing it absorbed heat. The hot vapor was recompressed outside the box, turning the vapor back into liquid and dumping the heat of compression away from the box's innards. The liquid was then pumped through the tubes back into the insulated box, where it vaporized again. In effect, the heat was pumped from the confines of the insulated box. In the early years, ammonia was used as the refrigerant, turning from liquid to gas and gas to liquid and back again in an endless cycle of compression and vaporization. Methyl chloride, sulfur dioxide, and carbon dioxide were also used. All of these were dangerous. As late as the 1920s, homeowners were killed by methyl chloride leaks. In 1928, General Motors asked a man named Thomas Midgley to find a better refrigerant, something nontoxic, nonflammable, and stable. It took Midgley and his team three days to come up with Freon, which is still used today in refrigerators built before the mid-1990s. Though slow to dominate the market, the refrigerator eventually killed the ice trade. And it was this principle — the vaporization of liquids to remove heat — that allowed very rapid but dangerous progress in the scientific exploration of absolute zero.

❄   ❄   ❄

It is February eighth and just below freezing in London. Fat, wet snowflakes fall lazily along the train tracks. Passengers are soaked from the knees down, dark trousers and skirt hems sticking to skinny legs, but the train is well heated. School has been canceled, and kids play in the open spaces between rows of small brick homes. There are few sleds but many snowballs. In a soccer field, a boy defends a pathetic snow fort, its walls melting around him, more slush than snow. Where he has harvested snow for the walls, a moat of still-green grass surrounds the fort. The boy is inexplicably shirtless, with shadows of his ribs showing on pale English skin.

The trains are running even later than usual. "I couldn't even get me car up the drive," a woman tells me.

Later, I walk through patches of sidewalk slush from Waterloo station to Westminster Abbey. Here, on a summer day in 1620, cold was an issue of some importance. This was two centuries before Frederic Tudor's ice trade and Thomas Moore's fur-lined icebox and Jacob Perkins's refrigerator. King James I—fifty-four years old, barrel-chested but somewhat bent over with rickets, a child of the Little Ice Age—did not do well in the heat. He overdressed, in part because he was a slave to fashion but in part to repel the knives of would-be assassins. Beneath his royal clothes, he tended to sweat. His skin itched. It is said that he became overheated when exposed to the sun.

When gray-bearded Cornelis Drebbel told King James that he could cool the interior of Westminster Abbey, the king listened. Drebbel, a Dutchman, was part scientist, part alchemist, part showman, part con man. He bragged of being able to change his appearance from one second to the next, of summoning ghosts, and of having created a perpetual motion machine. In Holland, he was known as the *pochans* or *grote ezel,* the braggart or big donkey, and he had been imprisoned in Prague for a combination of bad politics and bad debt. Now he lived through the largesse of King James. In exchange, on this day he would cool the interior of Westminster Abbey, the length of a football field with ten stories of open space between ceiling and floor.

The summer heat would have warmed the abbey's stone blocks. Even during the day, candles and perhaps lanterns would have burned inside, lighting the shadows and further heating the interior. This would not be an easy space to cool. But even an incremental cooling would impress an audience unaccustomed to air-conditioning. And although the room was tall, its air was still, and Drebbel would have known that the cold air—his cold air—would tend to stay low, near the floor.

There would be those who would see a change in the temperature as an act of magic, of sorcery, a summoning of winter in the middle of summer. Drebbel would do little to discourage such impressions. This was a time when cold was believed to come from a single source, called a *primum frigidum*. Aristotle himself believed that the *primum frigidum* was simply water, and in Drebbel's time Aristotle and the other ancient thinkers were still considered authoritative. Drebbel worked a hundred years before Fahrenheit and forty-five years before Robert Boyle's extensive work on cold, heat, and pressure. This was a time when controlled experiments and open communication about those experiments were not expected, when curiosity was by no means a virtue, when Francis Bacon was still formulating and promoting what would come to be called the scientific method. Neither the scientists nor their audiences were interested in sharing knowledge. The interest was in entertaining and being entertained, in amazing and being amazed.

Flash forward nearly four centuries. A verger shows a group of tourists around the abbey. I tag along asking questions. The verger is olive-skinned, wearing a black robe, his voice musical and his words and sentences made by combining clearly clipped syllables. He uses the word "chaps" in reference to long-dead royalty and even saints. He waves a flag to guide us through the abbey. He tells us that his title, verger, comes from the Latin for staff or rod. A verge is used by the verger to prod common worshippers away, allowing free passage for God's more important servants, for royalty and clergy and other favored mortals. He is animated, like a nervous little bird trying to stay warm, surrounded by the acid-worn stone figures of kings and queens and writers and artists and scientists. He shows us the verge that he uses, a brass baton, more symbolic than effective, the character of the stick perhaps reflecting the character of the man's position.

King James, no longer in need of air-conditioning, lies buried under the floor, the spot marked by a memorial tile. Laurence

Olivier's ashes reside here, too. Chaucer, or the body of someone believed to have been Chaucer, rests here, along with Dickens. Shakespeare did not want to be buried here but is honored by a stone figure. Close to Shakespeare's memorial, a wall tile mentions Mary Shelley. There is a tile, too, for Lord Kelvin, whose temperature scale went to absolute zero. And here is one for Faraday, a man who showed that melting ice absorbed heat — that the change from solid to liquid, and by extension from liquid to gas, absorbed heat in a way that could not be explained by the temperature change alone. Mix a pound of boiling water with a pound of water just above the freezing point, and you get two pounds of water at about 120 degrees. But mix a pound of boiling water with a pound of ice, and you get two pounds of water at something like 50 degrees. The change in state from ice to water — the breaking up of the molecular rows and columns that give ice its structure — accounts for seventy degrees of temperature change.

Viewed from above, Westminster Abbey has the shape of a cross. Drebbel likely emptied his bag of tricks in and around the Abbey's sacrarium, near the top of the cross. The walls and ceilings at that time would have been stained with soot and candle grease. Against these walls, Drebbel would have laid out casks or troughs. He had access to snow and ice stored in pits beneath nearby estates. He could find reasonably cool water just outside, in the Thames. And, importantly, he had potassium nitrate, also called saltpeter or niter. Drebbel knew, possibly from work first published in 1558 by Giambattista della Porta under the title *Natural Magick*, that mixing snow with niter resulted in sudden cooling. He was too secretive to leave written records, but Francis Bacon, who was not present that day in 1620, heard of the events. Bacon wrote that in "the late experiment of artificial freezing, salt is discovered to have great powers of condensing," and that "nitre (or rather its spirit) is very cold, and hence nitre or salt when added to snow or ice intensifies the cold of the latter, the nitre by adding to its own cold, but the salt by supplying activity to the cold of the snow." In short, niter mixed with ice yields an endothermic reaction. More commonly experienced exothermic reactions—such as the burning of wood or coal or gunpowder— generate heat, but endothermic reactions—such as the mixing of niter and ice—absorb heat.

King James and his entourage would have marched into the abbey, their gowns damp with sweat. They would have chilled quickly. Drebbel very probably added certain theatrical effects. The royal personage was entertained. His entourage was entertained. Some in the king's company may have feared this unseasonable cold. The king, who himself had once authored a book on witchcraft titled *Daemonologie*, likely exchanged clever comments with those standing nearby. And then they left.

The verger has never heard of Drebbel. Westminster Abbey remains today without air-conditioning. "It gets quite hot in the summer," the verger tells me. "Quite hot indeed. The stone walls themselves get hot, you see, so it doesn't cool off much in the eve-ning." Outside, the sun hangs low in the overcast sky. Here in the city, most of the snow is gone, but the sidewalks remain slushy. Lon-don is dreary and cold, King James I is long dead, and the Dutch-man Cornelis Drebbel is all but forgotten.

❄ ❄ ❄

Seventeen years before Daniel Fahrenheit came up with his tem-perature scale, at a time when neither molecular motion nor the atomic notion of matter were well understood, Guillaume Amon-tons reasoned that cold must bottom out. His reasoning was based on thoughts about changes in pressure and volume, then known to be affected by temperature. As it grew colder, pressure and volume decreased. Taken far enough, pressure and volume would have to reach into negative values, but negative pressure and negative vol-ume made no sense, so temperature, he believed, must have a bottom limit, an absolute zero. His thoughts made others realize that they lived in a very warm world, a world that hovered around the freezing point of water but that could be much colder. William Thomson, who would be knighted and known to posterity as Lord Kelvin, developed the Kelvin scale in 1848. The Kelvin scale used the same increments promoted by Anders Celsius, but it started with zero as the coldest possible temperature and put the freezing of water at a balmy 273 K.

Two centuries after Drebbel, before Kelvin developed his scale, scientists were reaching for this zero, turning their creative and intellectual talents toward what became at times an ugly competition for extreme cold. They took risks. They used chemicals that could burn one's skin and worked at temperatures that would crack both glass and metal. Things exploded. It was not uncommon for scientists to task their assistants with the hands-on work of the more dangerous experiments. Michael Faraday was the first to liquefy chlorine gas, in 1823, at a temperature of 130 degrees below zero. That same year, in the course of a single month, he was on the receiving end of three laboratory explosions, each of them causing minor eye injuries. Charles Saint-Ange Thilorier, the first to freeze carbon dioxide into dry ice, ran an experiment that resulted in an assistant losing both legs. In 1886, James Dewar's laboratory went up. Dewar himself, who had invented the vacuum bottle that was originally used in the laboratory and only later adapted for drinks, was nearly killed.

The scientists did not share the commercial interests of the Ice King, Frederic Tudor. Instead, they saw themselves engaged in a search for "the cold pole" and as surveyors working on "the map of Frigor." They saw their quest as equivalent to those of geographical explorers, of Columbus and Magellan and Cook, and of their contemporaries, the polar explorers Franklin and Greely and Scott. Medieval mapmakers had once labeled unknown lands in the far north "ultima Thule," and the scientists adopted the phrase. Heike Kamerlingh Onnes, who would liquefy helium in 1908, wrote, "The arctic regions in physics incite the experimenter as the extreme north and south incite the discoverer." He was right in this assessment.

Like the Arctic explorers, the scientists were men obsessed. They often invested their own wealth. They gave up opportunities for personal financial gain in exchange for opportunities to explore the frontiers of Frigor. They seemed to worry more about an experiment failing than about the safety risk posed by laboratory explosions and fires. Onnes, in a letter to Dewar, was concerned that "the bursting of the vacuum glasses during the experiment would not only be a most unpleasant incident, but might at the same time annihilate the work of many months." They had bitter rivalries, arguing over who reached the record low temperatures first, who was first to liquefy this gas or that gas. Frustrated when he realized that Onnes had won the race to liquefy helium, Dewar dressed down a senior assistant, blaming him for delays. The assistant vowed that he would never set foot in the Royal Institution as long as Dewar lived and then held true to his word.

These men, these explorers of Frigor, worked brutally long hours, conducting single experiments that could run from before dawn to late into the night, with laboratory assistants scurrying about, checking gauges, turning wheels that in turn spun threaded bars that in turn compressed gases. They opened valves, and—when things went well, when nothing broke or jammed or exploded—they transferred fluids such as liquid nitrogen, liquid oxygen, and liquid hydrogen from one vessel to another.

They were not above theatrics. In 1899, celebrating the hundredth anniversary of the Royal Institution, Dewar lectured to an audience of dignitaries and scientists. The men wore frock coats, and the women wore formal dresses. Dewar played with liquid oxygen. The men and women watched thermometers scream downward. Dewar liquefied oxygen before their eyes, turning a clear gas into a blue liquid. He showed them how electrical resistance decreased as metals were cooled. There were lots of tricks to be played near ultima Thule. Mercuric oxide went from scarlet to light orange. Rubber, ivory, feathers, and sponges phosphoresced with their own bluish glow.

The scientists expected even stranger phenomena as temperatures dropped further. Dewar himself, speaking of absolute zero, said that "molecular motion would probably cease, and what might be called the death of matter would ensue."

But absolute zero was unattainable. Reaching absolute zero can be likened to reaching the speed of light: as one approaches light speed, acceleration becomes increasingly difficult, and as one approaches absolute zero, further cooling becomes increasingly difficult. A simpleminded and not altogether wrong explanation: at very low temperatures, any attempt to remove more heat generates heat. But science is closer to absolute zero than to the speed of light. Science is within billionths of a degree of absolute zero, within spitting distance of ultima Thule. Things have become increasingly difficult and increasingly frustrating, and the death of matter, in a sense, has ensued.

If one ignores Drebbel's stunt at Westminster Abbey, the race toward absolute zero started in earnest in 1748, when the Scottish medical professor William Cullen used a vacuum pump to suck down the pressure of one vessel, cooling it sufficiently to freeze the water in a surrounding, outer vessel. He published the work in a Scottish journal under the title "Of the Cold Produced by Evaporating Fluids and of Some Other Means of Producing Cold," but he took it no further. Almost a hundred years later, cascading was developed, in which vaporization of one liquid cools another gas sufficiently to create a second liquid, which is in turn vaporized to cool another gas into a liquid state, and so forth and so on. Another trick involved the expansion of gas through a valve, allowing gas molecules to spread out without performing work and therefore to cool, taking advantage of what physicists know as the Joule-Thomson effect. At this point, still well above absolute zero, the men exploring toward ultima Thule were working at dangerously low temperatures, in a realm that left metal brittle and instantly froze flesh.

Helium was an important prize. In July 1908, starting before dawn, Heike Kamerlingh Onnes cascaded chloromethane to liquefy ethylene, liquid ethylene to liquefy oxygen, liquid oxygen to liquefy air, and liquid air to liquefy hydrogen. It had taken him seven years to prepare for the experiment. At seven o'clock in the evening, thirteen hours after the experiment began, Onnes became the first human being to see liquid helium. "It was a wonderful moment," he later recalled. Helium goes from gas to liquid at 452 degrees below zero Fahrenheit, less than seven degrees above absolute zero.

Below the temperature of liquid helium, cascading and the Joule-Thomson effect are of little value. Things become increasingly peculiar. For Onnes and his colleagues, trained in classical physics, the properties of matter no longer made sense. Even calling liquid helium a liquid is not quite right. It is more of a superfluid, a phase of matter that behaves something like a liquid but that has almost no viscosity.

Albert Einstein weighed in. Working with the Indian physicist Satyendra Nath Bose, Einstein realized that quantum physics was at play. In the quantum world, atomic motion occurs in incremental steps: it is as though you can travel at one mile per hour, five miles per hour, or ten miles per hour, but not at three miles per hour or four and a half miles per hour or seven and a quarter miles per hour. Bose and Einstein realized that at extremely low temperatures, the wave functions that described individual atoms would overlap. The wave functions would then merge, and groups of atoms would behave as one. In 1924, Bose and Einstein proposed that a new state of matter would exist at extremely low temperatures. This state of matter—not gas, not liquid, not solid—became known as the Bose-Einstein condensate.

It took sixty years to develop the technology to knock off those last couple of degrees on the road to the Bose-Einstein condensate. By then, both Bose and Einstein were dead. In the quest for ultima Thule, new tricks had been discovered. When light hits an object and is absorbed, the object is warmed, but when light hits an object and is reflected, it is cooled. With lasers, pure wavelengths can be generated and reflected off a packet of atoms. In 1995, Eric Cornell and Carl Wieman used lasers to cool a packet of rubidium atoms to ten-thousandths of a degree above absolute zero, to the point at which molecular motion almost stops. "It's like running in a hail storm so that no matter what direction you run the hail is always hitting you in the face," Wieman said. "So you stop." It was still too hot for a Bose-Einstein condensate. To cool their rubidium atoms further, they used what in principle is the same trick used by their nineteenth-century forebears—a form of evaporative cooling, allowing the most energetic atoms to escape and leaving only the less energetic and cooler atoms behind. "It's the exact same physics as how a cup of coffee cools," Wieman explained during a 2001 press conference. "The steam coming off is the most energetic coffee atoms. The ones left behind get colder." In the end, the rubidium had cooled to something like fifty-billionths of a degree above absolute zero, or just shy of 460 degrees below zero Fahrenheit. And the atoms were no longer a gas or a liquid or a solid. In a brand of alchemy that would have made the likes of Cornelis Drebbel cackle in glee, the kind of matter known to men such as Dewar and Onnes and Boyle had died, and what was left was a Bose-Einstein condensate—a thick glob of about two thousand atoms condensed into a single super atom surrounded by warmer atoms, reportedly looking something like the pit inside a translucent cherry made of a glowing cloud of very cold rubidium.

❄   ❄   ❄

It is February twentieth and forty-one below in Fairbanks—419 degrees Fahrenheit above absolute zero. True to form, the cold has returned to Alaska. Frigid air hitting cold lungs makes me cough. My parka is stiff. Its fabric complains when I move, screeching softly, retaining wrinkles, holding its shape, its molecules locked up as if close to ultima Thule. My rental car was parked outside overnight, and its tires have frozen out of round. As they thump down the road, I bounce up and down as if driving a Mack truck. Outside, I pull on my hood. I hunch my shoulders and walk with my head down. Today, forty-one below leaves me grumpy and withdrawn.

In my hotel, the International Arctic Research Center has teamed up with a consortium of Japanese universities to hold a conference on global warming. The language barrier makes conversation all but impossible. From what I can gather, there is concern that melting permafrost will crack foundations and cause sags in oil pipelines. A German meteorologist argues with a Japanese glaciologist in three languages. They wear name tags in plastic holders strung from their necks. Within this meeting room, to a degree seldom seen elsewhere, there is a realization of the extent of the world's permafrost and the threat of a warming planet. There is a realization that permafrost will melt in Alaska and Canada and northern Russia and Norway and in the mountains of Tibet. The same warming that will compromise a trapper's cabin in the Alaskan bush will fracture a thousand-year-old Buddhist temple in the Himalayas.

No one goes outside. For scientists, it is too cold to go outside.

Front-page news in Fairbanks focuses on the thousand-mile Yukon Quest dogsled race. Lance Mackey, having made it to the Chena Hot Springs checkpoint late yesterday afternoon, is set to break the record. After a required rest stop, he has but ninety-nine miles to go. He will likely mush into Fairbanks sometime early this afternoon, with a race time of just over ten days, beating the 1995 record by half a day. These are tough races for both dogs and mushers. One of the athletes, a sled dog named Melville, died close to the Yukon River. Another, named Jewel, died earlier in the race, choking on its own vomit. But the mushers do not win races by merely riding along behind the dogs. The sleds have to be worked through drifts. They pound over rough snow and ice. These are the trails of prospectors and trappers and Jack London, and before them of the Athabascan Indians, who lived inland from the Eskimos, in Alaska's brutally cold interior. On steep runs, it is not unusual for a musher to run along behind the sled. Along the Yukon, they may run along in temperatures that hit seventy below, plus windchill. Photographs of the mushers show their faces iced over, looking drawn, caring for one dog or another, hugging it or feeding it or working the harness straps under the light of a headlamp.

On a back page, the Fairbanks paper talks of three climbers and their dog saved on Mount Hood, Oregon, hauled away in an ambulance after spending the night huddled in the snow, like maladapted penguins. "We're soaking wet and freezing," one of the climbers said. According to one of the rescuers, the dog, lying across their bodies through the cold night, likely saved their lives.

On the East Coast, discount air carrier JetBlue has just recovered, back on schedule a week after a snowstorm forced it to cancel 139 flights. Closer to home, in Soldotna, a couple of hours south of Anchorage, officials estimated damage to city property at more than one and a half million dollars, all of it linked to the short-lived but very real warm spell of the week before, with its quick thaw causing flooding and the moving floodwaters carrying big blocks of ice, reminding everyone of the temporary nature of human structures in Alaska.

And there is this in the newspaper: human egg donors are getting as much as eight thousand dollars from prospective parents. More than ten thousand women donated eggs in one year. They used the money for vacations and college tuition and rent. Men, too, are part of this economy. "Our main activity," one company advertises, "is worldwide delivery of high quality frozen tested semen from more than 250 donors." Sperm are stored in liquid nitrogen at 320 degrees below zero. Once stored, sperm can be kept indefinitely. It is reasonable to expect fifty percent survival even after hundreds of years.

For eggs, freezing is far more risky, but when women are threatened with various diseases that would otherwise end their ability to reproduce, freezing can be an option. Reluctant parents could, in theory, freeze sperm and eggs for generations, have the sperm and eggs joined long after they themselves are dead, and reproduce without the headaches of child rearing and babysitting. Perhaps Fahrenheit and Boyle and Faraday would be mortified to know of where their research has led. Alternatively, perhaps they would have masturbated into test tubes and stored their sperm in liquid nitrogen, to be resurrected as needed, little half geniuses waiting centuries for a nice warm egg to inseminate.

❄ ❄ ❄

What good can come from a couple of thousand molecules of rubidium at 460 degrees below zero? Of what use is a Bose-Einstein condensate?

For one thing, very cold temperatures win Nobel Prizes. Onnes got one in 1913. The discoverer of a method of cooling called adiabatic demagnetization got one in 1949. The trio of physicists who developed laser-based cooling methods got one in 1997. And Eric Cornell and Carl Wieman received one in 2001 for their work with rubidium atoms near absolute zero. At a press conference, Cornell was asked how it felt. "Pretty good," he said.

Drebbel's air-conditioning of Westminster Abbey came to nothing. To the extent that he is remembered at all, it is for a submarine that he plied along the Thames River. He died poor. William Cullen's first refrigerator in 1748 was little more than a footnote. Cullen himself did not see any commercial value in refrigeration. John Gorrie, working on both refrigeration and air-conditioning, and actually using it to cool both his home and a hospital ward, died in 1855 before he could attract investors. Potential investors, it is said, believed that it would be cheaper to ship ice south than to make it in Florida. Another ice plant went belly-up when its owner made the mistake of building it in Minnesota. But in 1859, Ferdinand Carré sold a cooling machine to a Marseille brewery. By 1889, cold air was being piped from central cooling plants to customers in New York, Boston, Los Angeles, Kansas City, and St. Louis. And by 1902, the New York Stock Exchange was air-conditioned.

By 1880, refrigeration was cheaper than natural ice. With efficient refrigeration, food could be shipped to markets. Hog production grew. Over a few years, beef exports from the United States to the British Isles grew from a hundred thousand pounds a year to seventy-two million pounds a year. More than a hundred thousand refrigerated railroad cars appeared almost overnight. Suddenly, midwestern farmers could undercut New England farmers on dairy products. Refrigeration increased the profitability of ranching, and ranches expanded, implicating refrigeration in the last phase of the eviction of Native Americans from their ancestral lands and the near extinction of the buffalo. Clarence Birdseye, inspired by the quality of fish frozen at forty below after ice fishing, developed techniques for flash freezing food in 1923. By 1928, Americans were buying a million pounds of flash-frozen food a year. With air-conditioning, people could live comfortably in Florida and Arizona and New Mexico. By virtue of the air conditioner, skyscrapers grew taller, into the range of gusting winds that prevented the opening of windows.

At colder temperatures — the temperatures of Onnes and Dewar, the temperatures of liquid nitrogen and liquid oxygen — steel manufacturing was improved. By 1914, Robert Goddard was using liquefied gases as rocket fuel. The Mercury rockets, the Gemini rockets, the Apollo rockets, and the space shuttle have all used liquefied gases for fuel. The tip of a probe threaded up a patient's femoral artery to the heart and chilled to a hundred below can treat cardiac arrhythmias. At 180 degrees below zero, old tires shatter, allowing their material to be recycled. Two hundred and fifty pounds of superconductors at the temperature of liquid nitrogen can replace eighteen thousand pounds of copper wire. Cryosurgery, cryogenic tire recycling, the liquiefication of natural gas, and the manufacture of sophisticated electronic equipment require temperatures that could not be reached if not for the obsessions of Onnes and Dewar and their colleagues and competitors, all of whom, at one time or another, were likely asked about the usefulness of their explorations of Frigor.

In 1997, researchers came up with a way of dripping single, very cold atoms from a microspout. In 1999, it became possible to fire streams of atoms in any direction, a kind of atomic laser. Also in 1999, a Harvard University team shined light through a Bose-Einstein condensate, slowing the beam to just thirty-eight miles per hour. In 2001, a beam of light was, for a moment, stopped entirely, and by 2007 the team could slow light down, stop it, and restart it. The team leader, Lene Vestergaard Hau, told reporters, "It's like a little magic trick." The secret behind the magic: a tiny lump of Bose-Einstein condensate.

Useless stuff, all of it, surely no better than the cooling of Westminster Abbey or Dewar's 1899 lecture-hall tricks with liquid hydrogen.

MARCH

It is March fifth and five degrees below zero in Anchorage. Throughout the day, I have been indoors and warm, taking a winter survival course with a crew of biologists headed north to look for polar bear dens. At their research site, on the edge of the Beaufort Sea, the temperature hovers around forty-seven below zero, not counting windchill.

The instructor shows us his parka, insulated with a hollow synthetic fiber. He shows us how the hood sticks out past his face, forming a snorkel that tunnels heat. "Twenty-five percent of radiant heat loss," he says, "is from your head."

But you are leaking heat everywhere. You sweat, and every molecule that changes from liquid to vapor takes away heat. You sit on the ground, and heat is conducted from your rump to the snow. You stand, and air moves past your body, even on a still day, the convection currents bringing in cold and taking away heat. You breathe, sucking in cold air and exhaling warmth. You are a human radiator giving up the heat you need to stay alive.

The instructor talks about hypothermia. "You're not dead until you're warm and dead," he says. He shows a frostbite video: lifeless fingers blackened to the knuckles.

He talks about the importance of a trip plan. "Know where you are going," he says, "and make sure someone knows when you will be back."

"Do you have things on your person that will help you survive?" he asks.

"Cotton kills," he says, discouraging us from jeans and cheap T-shirts. He urges us toward certain synthetic fibers.

He talks about hypothermia warning signs, which he calls "the umbles." You mumble as your jaw muscles chill. You fumble as your fingers stiffen. You grumble complaints as your core temperature drops. You stumble drunkenly as your central nervous system slows. You tumble. You are down in the snow. Without help, you will die.

"You're not dead," someone quips, "until you're warm and dead."

After the class, I make the mistake of handling my skis and poles with bare hands. Chilled metal of ski edges and poles sucks away heat. I pull on light synthetic mittens. They are ten years old, and the palms are worn thin. An index finger, numb, shows through a hole. Some sort of white stuffing, a petroleum product of some kind, pokes out at a seam. I ski hard under birch trees in the flat evening light, but ten minutes later both hands are numb. I rip open two chemical heat packs, one for each glove. They are all I have on my person. Each pack is two inches on a side, a thin white mesh holding a black powder, nothing more than iron, salt, some wood fiber, and activated charcoal. Exposed to air, a chemical reaction moves forward, and the pack releases heat. Their quality varies. At five below, with holes in my mittens, it is hard to tell if they are even working.

My dog runs into the forest and reappears a moment later, the frozen carcass of a snowshoe hare clutched in his jaws. With his trophy he trots along behind me, without a trip plan, with no under-standing of frostbite, which affects dogs just as it affects people, with nothing but the rabbit on his person to help him survive, wearing neither synthetic fibers nor cotton, tail arched upward, ears erect, head up. If dogs can smile, he is smiling now.

❄ ❄ ❄

Angora rabbits, highly bred for long soft fur, look more like puffballs than rabbits. They were first raised in the Carpathian Mountains near Poland when the local people realized that rabbit fur provided a softer alternative to sheep's wool. This occurred around the sixth century, about the same time the spinning wheel appeared in India and several hundred years before cotton socks showed up in Egypt. The word "angora" was used in the language of these people and is sometimes said to mean "not sharp."

Sheep's wool feels coarser than angora, scratchier, but also warmer. Sheep, too, are easier to care for than rabbits and far easier to herd. Wool was being spun into yarn and used in clothes for at least four thousand years before anyone tried the same trick with rabbit fur. Today there are more than forty breeds of sheep produc-ing two hundred varieties of yarn. There are the soft wools, merino and rambouillet and debouillet. There are the medium wools, such as that of the Finnsheep, said to be good for gloves and sweaters and jackets. There is the coarse wool of the Scottish Blackface and the Welsh Mountain sheep, the latter said to be good for rugs.

Wool is judged by the diameter and length of its individual hairs, as well as by color and curl. The cream of the wool crop comes from the sides and shoulders of the sheep. After shearing and sorting, grit and dirt are washed from the wool, and then it is passed through rollers with wire teeth in a process called carding. Carding untan-gles the fibers and arranges them into a flat sheet called a web. From there, the wool becomes yarn of various diameters, and the yarn becomes sweaters or pants or shirts or rugs, all with a quality that is distinctly absent from cotton. Wool, when wet, stays reasonably warm, losing only a fraction of its insulative character, but cotton, when wet, conducts heat nine times more efficiently than it does when dry. Take a pair of cotton jeans, throw them in a mountain stream or a pond or a bathtub full of water, and take them out. They are heavy with water. Cotton kills because the fibers hold water. They will suck the heat right out of the unfortunate victim foolish enough to wear them in the cold. For survival in the cold, naked skin may be better than wet cotton jeans. Take a pair of wool pants, throw them in water, pull them out, and they are not so heavy. Wool, under a microscope, is oval in cross section. It curls as it grows and is naturally springlike. Each hair is covered with scales and coated with a waxy outer layer of lanolin that repels water. Air spaces reside between the curls, even in wet wool, and it is these air spaces that insulate the wearer.

The key to warmth is to make sure that air does not flow. The key is to trap air in small spaces, in the spaces between the curls of wool fibers or between layers of clothing.

Toward the end of World War II, before the Americans dropped two atomic bombs on Japan, preparations were under way for what promised to be a very ugly amphibious invasion. The U.S. Army's Wet-Cold Clothing Team was deployed to Kwajalein, Saipan, Tinian, Guam, and Iwo Jima to teach the principles of layering to men accustomed to fighting in tropical jungles. In a jungle clearing or just before a USO show or in a mess tent, the Wet-Cold Clothing Team demonstrated the use of wool clothes. "Keep it clean," they would tell the men. "Keep it dry and wear it loose." They described heavy losses from hypothermia and frostbite in the war in Europe. They described the permanently crippling effects of frozen fingers, toes, and feet. Under the tropical sun, the instructors explained that staying warm depended on the trapped air between layers that were individually reasonably light. Part of the trick was to wear what you needed and carry the rest, thereby avoiding sweating. Put on a layer when you are cold, and take off a layer when you are warm. The instructors would put on a layer of wool, explain its purpose, add another layer, explain its purpose, add another layer, explain its purpose, and so on, until they were fully layered for cold-weather soldiering. The training likely helped thousands of soldiers who had never before been confronted by cold. The heavier instructors lost twenty pounds donning woolen layers in the tropical heat.

Cashmere wool comes from the Indian cashmere goat, and mohair wool is from a North African goat. There is alpaca wool, vicuña wool, and llama wool. In Alaska and parts of Canada, there is musk ox wool. When warmth and cost are considered together, sheep's wool is the best of the natural fibers. When warmth alone is considered, musk ox wool wins, reputedly the warmest fabric known. It is as much as eight times warmer than sheep's wool, and it is softer.

An adult male musk ox weighs something like eight hundred pounds. It stands a bit over four feet tall at the shoulders. Its expression, at best, is somber. The bulk of its shoulders and the curl of its horns, which sweep downward from in front of its ears and then upward and outward, resembling a deep scoop, make the animal appear hunched and somehow grossly unintelligent. In the wild, a threatened herd of fifteen or twenty musk oxen will form a circle, allowing a man with a rifle to shoot one after another. Admiral Peary's expedition to the North Pole killed something like six hundred of them for food. But the oxen are not killed to make wool. Instead, they are raised in captivity, and their *quviut*, or underfur—the soft warm stuff that hides under their bristly guard hairs, the mammalian equivalent of the down found under the coarser outer feathers of birds—is combed out each spring.

A full-grown musk ox naturally sheds something like five pounds of quviut each year. The stuff is remarkably light. Five pounds, stuffed and packed, will fill a kitchen-size garbage bag. Oomingmak, the Musk Ox Producers' Co-operative, buys all the fiber they can find. If they can pull together six hundred pounds—enough to fill 120 kitchen-size garbage bags—they ship it to a company on the East Coast, where it is treated in much the same way as sheep's wool and spun into yarn. Months later, sometimes more than a year later, the yarn is distributed to knitters, typically Native American women working in villages scattered around Alaska. At their own pace, they knit scarves and hats and tubular pullovers called smoke rings, which can be used to warm the neck, be pulled over the head and face like a balaclava, or be worn on top of the head like a chimney hat. Marketers claim that quviut yarn is as soft as a cloud. They claim that a woman with her eyes closed could touch this stuff and never feel it. A quviut cap of the kind one might wear skiing costs more than four hundred dollars. Quviut scarves go for about the same price as hats. Quviut smoke rings are a bargain at under three hundred dollars.

For warmth, animal fibers outdo plant fibers, but plant fibers cost less. Picking cotton may be backbreaking work, but it is easier than combing quviut from under the belly of a musk ox or raising puffball Angora rabbits or even shearing sheep. And it was plant fibers that led to synthetics. Rayon was inspired by silk—which then and now comes from a caterpillar's backside—and was commercialized in the late 1800s. Count Hilaire de Chardonnet collected wood chips and treated them chemically to get a substance called viscose. The viscose was pushed through a spinneret, which looks and functions something like a showerhead, but instead of sprinkling threads of water, it sprinkles threads of fiber. The sizes of the holes in the spinneret determine the thickness of the fibers. The count took his product to the 1889 Paris Exhibition and two years later started manufacturing and distributing large quantities of viscose at a factory in France.

Rayon was the first manufactured fiber, but it is not a true synthetic. Rayon comes from wood fibers, while true synthetics come from materials that do not even resemble fibers, things such as amine and hexamethylenediamine and natural gas and oil. Nylon was invented by Wallace Carothers of Du Pont, who had once taught organic chemistry and who specialized in the study of long chains of repeating units of atoms called polymers. By 1935, his team had more or less perfected a method of combining amine, hexamethylenediamine, and adipic acid in a way that would yield molecules with more than one hundred repeating units of carbon, hydrogen, and oxygen. A filament of nylon might have a million of these molecules. Nylon was first used commercially for toothbrush bristles. By 1939, nylon was also used in stockings, fishing line, parachutes, and lingerie. By 1945, people were fighting over the stuff. On more than one occasion, police were summoned. In the worst of what have been called the Nylon Riots, forty thousand women fought over thirteen thousand pairs of nylon stockings in Pittsburgh.

By the 1990s, a confusing array of fabrics and advice were available. Nylon was still around, good for blocking wind. And there was polyester, good for insulation even when moist, but heavy and clammy when wet. There was polypropylene, which moves water away from the skin. There was Polarguard, a single strand of a polyester-like material widely used as sleeping bag insulation—warm when wet but relatively heavy. There was Hollofil, with hollow fibers, and Quallofil, like Hollofil but with four holes through the fibers. There were the breathable fabrics—Gore-Tex and Stormshed and Klimate—with pores a few hundred times bigger than a water molecule. Water vapor can escape through the pores, but anything as big as a raindrop cannot squeeze through. There was PrimaLoft and Micro-loft, superthin fibers that are nearly as light as down but that do not form clumps when wet, as down does, making them warmer than down when wet.

There are ideas, too, that have not yet been and may never be widely used in clothing. There was the idea that sealing moisture in might be beneficial under the right circumstances. A vapor barrier against the skin would signal the body to stop producing its base level of perspiration. Because trapped moisture cannot steal heat by evaporating, a camper sealed within a vapor barrier might be warmer than one whose clothes breathe. On the downside, the system fails if the camper overheats from hiking or cutting wood or running from a bear, producing pools of sweat that have nowhere to go. Also, few campers are comfortable lying in their own steam, and at the end of a three-day trip, they tend to stink.

There was the invention and patenting of heated clothes. There are socks with built-in pockets for chemical heat packs, such as the "Thermal sock having a toe heating pocket," described in patent number 5230333, issued in 1993 to James and Ronnie Yates. There is the "Electrically heated boot sock and battery supporting pouch therefore," patent number 3663796, issued in 1972, the "Inflatable boot liner with electrical generator and heater," patent number 4845338, issued in 1989, and the "Electrically heated garment," patent number 5032705, issued in 1991.

And there are the so-called smart fabrics and smart clothes. Some have porosity characteristics that change with temperature, the pores growing to release heat and moisture near hot patches of skin or shrinking to preserve heat when skin temperature drops. "It's very simple," one of the inventors reported. "You cut flaps in the clothing, and as the fabric absorbs water, one surface swells up and the flaps bend backwards."

Some even smarter fabrics have built-in microprocessors. In addition to keeping the wearer warm, the smarter fabrics have the potential to become wearable computers, providing navigation and communication aids and monitoring the wearer's pulse and breathing rate.

For his well-being when exploring the Arctic in the early 1900s, Vilhjalmur Stefansson consistently shunned the European ways and turned instead to traditional local ways. "For nine winters I have never frozen a finger or a toe nor has any member of my immediate party," he reported.

The Inuit, then and now, wear fur. On very cold days, one approach is to wear an inner layer with the fur turned toward the skin and an outer layer with the fur turned outward. Native American Niomi Panikpakuttuk said in a 1996 interview, "Of course caribou skin was the only source of clothing that we could get when I was young. The textiles that were available to us were not good for winter wear. As a matter of fact, I do not consider them to be the type of material that you could use in winter. I am still like that; whenever I am wearing textiles, I have to put on layers and layers of clothing on my body and legs, and even at that it will not warm me up. This is because I am a real Inuit. I do not consider textiles warm clothing."

In Anchorage today, two or three hundred dollars will buy oversize mittens of beaver, coyote, fox, or lynx. A fur hat can cost more than five hundred dollars. Mukluks, perhaps the ugliest boots ever made, cost four hundred dollars in either beaver or coyote. Bikinis made from lynx or fox or wolf, though of questionable value in the cold, might be considered a bargain at under three hundred dollars.

❄   ❄   ❄

It is March eighth and twenty-three degrees below zero here at the edge of the Beaufort Sea, reasonably warm for this time of year. On the East Coast, cold weather is in the news. York, Pennsylvania, is experiencing a record low for March, at minus nine degrees. Atlantic City, New Jersey, is also experiencing a record low, at four above. March is an unpredictable month. On this day in 1941, Philadelphia got eight inches of snow and at fourteen degrees experienced the second of three consecutive record-low-temperature days.

We have turned off our snowmobiles, silencing them, darkening their headlights. The wind blows. Above us, eighty miles up, the northern lights stretch in a pale green arc over the pack ice. A hundred feet in front of us, there is a low rise. Near the center of the rise, a polar bear den had been spotted in December. This is the den that we hope to find. The wind picks up snow between us and the rise, carrying it just over the ground, dusting our boots and forming tiny drifts against the brown musk ox droppings that lay scattered around our feet.

On my torso, I wear two polyester and spandex shirts, covered by a light nylon jacket stuffed with polyester PrimaLoft fill, all buried under a thick down parka intended for use at the South Pole. My hands nest inside loose-fitting gauntlet-style mittens. On my face, I wear a full mask, polyester and spandex, with cutouts for eyes and nostrils and a small one where my mouth should be. I also have my parka hood, which for the most part I leave in the down position. If I pull the hood up, it forms a snorkel in front of my face, muting the wind, and I have a tunnel-vision view through the coyote-fur ruff that muffles moving air six inches out. On my lower body, my feet rest inside lined boots. I wear polyester and spandex tights, a pair of looser pure polyester trousers, another pair of thicker but even looser polyester trousers, and down-filled bib overalls. The down has been treated to fight bacteria. The outside of the overalls has a Teflon fabric protector that repels water but is breathable. Underneath it all, I wear a pair of thick flannel boxer shorts printed with polar bears.

I feel as if I am wearing a space suit. I look like Charlie Brown dressed for winter. I am warm. Farther south, I would be grossly overdressed, but here I am stylish, in vogue.

In December, the polar bear had been spotted from the air with an infrared scanner. At that time, she was digging the den that we hope lies out in front of us. A video clip from the scanner shows the bear turning to look at the airplane, seeming to stare right into the scanner. Her teddy-bear ears glow with warmth. She moves about in her den, which is perhaps three times the size of her body. Her movements within the freshly dug den and her stares toward the airplane somehow speak of her solitude. But by now she should have cubs, little bears that are cuddly but no doubt annoying within the confines of the den. With a handheld infrared scope of the kind used on search-and-rescue missions, we search for her. We see nothing but the varied shades of gray and white that make up the snow-covered hill. We move closer but still find nothing. The bear, if here, is not giving away her position. Perhaps the den has become drifted in, the snow blocking the heat signature, or perhaps she left, cubless, or even with cubs in tow, earlier than expected.

I turn the infrared scope toward our snowmobiles. All three glow hot white against a background of gray snow. I turn the scope toward my companions. Their bodies are ghostly gray outlines with white blotches where they leak heat. They leak heat through their boots and under their arms and through their gloves. Their faces glow hot white, seventy-five percent of their heat loss purged out through their snorkel hoods.

One of the snowmobiles is dead. Its key start results in nothing but the useless whir of the starter, and its pull cord is jammed. We pop the hood. Inside, the bendix — the end of the starter motor that engages with the flywheel to start the engine — has broken off and wedged itself against the flywheel. One of my companions pulls it free, and I drop it into my pocket. "Nothing lasts at these temperatures," my companion says. Today he has broken the faceplate of his helmet, the copper wire that powers his helmet heater, and now the bendix. But with the bendix removed, we can start the machine with its pull cord. We head south, the noise of our machines silencing the wind, our headlights darkening the northern lights, our presence temporary, ephemeral leakers of heat fleeing south toward the warmth of a permanent camp powered by the almost bottomless pit of natural gas that resides far beneath the permafrost.

❄   ❄   ❄

For the denned polar bear, the outermost garment is not its fur but its den. For the human, the outermost garment is not the caribou coat or parka or Windbreaker but the igloo or snow cave or quinzhee or house.

Much has been made of the Inuit igloo. Otherwise reasonable adults have the misguided subliminal impression that modern Inuit live in igloos. The igloo today is at most a temporary shelter, sometimes used instead of a tent for a few nights while hunting or traveling. A skilled builder with suitable snow — snow reworked by wind that can be cut into blocks with a long-bladed snow knife — can build a small igloo in a few hours. The hole from which the snow blocks are cut forms the bottom half of the igloo. Around the hole, the blocks of snow are stacked, each one angled slightly and with their size growing progressively as the igloo walls grow. Envision a spiral of bricks growing larger with each step around the spiral. The warmth of occupants warms the interior. Over time, with melting and refreezing, the interior becomes slick ice. The blocks freeze into a single integrated structure. Fur can be hung on the walls for better insulation. The entrance is through a tunnel that dips down and then up into the dome. Envision crawling through the snorkel opening into a warm hood. The temperature outside might be forty below, while the temperature inside may be well above freezing.

In the past, igloos were larger and could be occupied for an entire winter. Igloo villages could be found scattered around the far north, especially in Greenland and the Canadian Arctic. In his 1932 publication *The Indians of Canada*, the Canadian anthropologist Diamond Jenness wrote about igloos:

> Glance for a moment at the interior of an ordinary, single-room snow hut. You pass with bowed head along a narrow, roofed passage of snow blocks until you arrive at the doorway, a hole at your feet, which you traverse on hands and knees. You rise to your feet. On the right (or left) two feet above the floor is the lamp, a saucer-shaped vessel of stone, filled with burning seal-oil, and with a stone cooking pot suspended

above it. Behind the lamp are some bags containing meat and blubber; in front of it, a wooden table bearing perhaps a knife and a ladle. A low platform covered with skins occupies fully half the floor space. There, side by side with their heads facing the door, the inmates sleep in bags or robes of caribou fur. If you stand at the edge of this platform, exactly in the centre of the hut, you can place both hands on the ceiling and almost touch the wall on either side. A thermometer three feet from the lamp will register one or 2 degrees below the freezing point of water, quite a comfortable temperature if you are enveloped like the Eskimo in soft, warm garments of caribou fur.

In Jenness's day, several igloos could be joined by their entrance tunnels or walled together. The largest could have five rooms, ice palaces in the northern wilderness, mansions of snow. Some communities built large snow domes for dancing and singing and wrestling. Certain communities are said to have settled disputes through singing contests in which the point of the song was to ridicule a rival. The songs could go on for hours, much like litigation today but more melodious and possibly as just.

In the summer, igloos melt. A point would come at which a tent was erected. The community was mobile, moving along the coast to hunt whales and caribou and fish. When asked if this seemed inconvenient, an Inuit might have responded that summer is the season for being outside. Why would one want a house in the summertime?

Away from the coast, in the taiga forests that cover hundreds of thousands of square miles, the snow is powdery and often sparse. It cannot be cut into blocks. The Athabascan people of interior Alaska built *quinzhees,* piles of snow that they hollowed out. The beauty of the quinzhee is that, unlike the igloo, any fool with a shovel and a bit of snow can build one. In fact, even the shovel is optional. The snow is scooped off the ground and mounded, then left to sit for an hour or two. The pores in the snow are saturated with water vapor. There are differences in temperature within the snow, and the water vapor moves from the warmest pore spaces, where vapor pressure is highest, to the coldest pore spaces, where vapor pressure is lowest. As the vapor moves, it cools, turning to liquid and then ice. The snow metamorphoses. It hardens. In two hours, it is hard enough to allow tunneling. The savvy tunneler pokes foot-long sticks into the mound, making it look something like a pincushion on the outside and providing guidance on just how much snow to shave away from the inside. The tunneler digs upward at a slight angle and then hollows out the dome. A bench might be left on one side. An airhole might be a good idea. A quinzhee, while lacking the elegance of a well-made igloo, can last through the winter.

Today the villages of northern Alaska rely on imported houses designed for temperate climates. The houses, or at least the materials from which they are made, travel north by barge. This is an expensive trip. After erection, decay begins almost immediately. Neither the materials nor the designs were meant for the Arctic. From the outside, wind carrying snow blasts the structure. The ground beneath, warmed by the structure itself, melts and subsides. From the inside, the humidity of human life—exhaled air, steam from a coffeepot or a shower, water evaporated from washing countertops or from houseplants—finds its way toward the walls and ultimately into the walls themselves. Just as water vapor moves through the pore space in snow, going from warmer pockets to colder pockets—from higher vapor pressure to lower vapor pressure—it moves through walls. It finds its way through any available opening. It moves through the insulation. Somewhere in the insulation, between the studs or joists, it condenses into liquid water and then freezes. In spring, it thaws. Water stains form on walls and ceilings. The water refreezes, and the force of expanding ice pops nails and tears screws from wood. Rot and mildew settle in. What was delivered by barge to become a fine little house quickly becomes a very expensive hovel. The mobility of the ancestral way of life looks increasingly attractive.

In Fairbanks, the Cold Climate Housing Research Center looks for ways to improve on the outermost garment. For something over five million dollars, the center provides fifteen thousand square feet of space for labs and offices, but more important, it is a living experiment in improved construction for use in cold climates. The foundation can be jacked up if the ground beneath starts to melt. After jacking, structural foam or concrete can be injected into the open space. The building's vapor barrier is outside its wooden frame but encased in a blanket of polystyrene, three layers thick, which itself is encased in stucco siding. What little vapor might escape through the barrier will not condense and freeze against the wooden frame of the building. Also, the building's ventilation is set up to remove moisture, to vent it to the outside, but for the most part the warm air is reused. New air — cold air from outside — is sucked in only as needed. Just as in a snow cave or a quinzhee, in a polar bear's den or a lemming's subnivean run, carbon dioxide can build up, so the building has carbon dioxide detectors that feed information to the ventilation system.

In the lobby, what looks at first like an ornate fireplace made from beach cobbles turns out to be a masonry heater. The flue snakes around and turns back on itself, an extravagant twisted snorkel hood of stone. A single morning fire, fed with perhaps thirty pounds of very dry wood and all the air that it can suck in from the surrounding room, burns hot and fast. The heat snaking around in the flue heats the rock. Throughout the day, the cobbles cool, dumping their heat into the building.

There are five hundred sensors embedded in the walls and floors and ceilings of the Cold Climate Housing Research Center, measuring temperature and humidity. There are sensors, too, in the ground beneath the foundation. To the extent that this place leaks heat, its operators will know. As an outermost garment, it is like smart clothes, with layers, or like a combination of layered smart clothes and the sort of rebreather mask worn by Richard Byrd, overwintering alone at sixty below in Antarctica. But not at all like Byrd, the people inside the building are hot. They shed their jackets like hikers removing a layer before they sweat. They sometimes open the windows of their offices, setting the heat free and letting in the cold and feeling suddenly alive in the middle of the business day.

APRIL

It is April second and well over sixty degrees in San Francisco. Outside my hotel, Californians mill about, some walking on the streets or hanging from streetcars that pass beneath my room or hidden in cars and buildings. There are more people in this city than in all of Alaska. How such crowding is tolerated baffles me. And yet their roads are not frost-heaved, the windshields of their cars are not cracked from the combination of flying gravel and bitter cold, and their pipes are not frozen.

Mark Twain worked here as a journalist. One of his editors is said to have claimed that Twain could not be trusted with anything but obituary notices and that he wrote them in advance, leaving a space for the name of the deceased and another space for the date of demise. A senator claimed that Twain offered to write favorable copy for him in exchange for an open bar tab. After a month, the senator stopped paying the bar tab, claiming that Twain had failed to produce and that the bar tab "twas very large." One can imagine Twain rationalizing his drinking as a means of bracing himself against the city's legendary cold. It is often claimed that Twain once said that the coldest winter he ever spent was summer in San Francisco. In fact, he never said that at all. The quote is one of many falsely attributed to Twain, including these: "There are three kinds of lies: lies, damn lies, and statistics," which Twain himself attributed to British prime minister Benjamin Disraeli; "Wagner's music is better than it sounds," which Twain attributed to the humorist Edgar Wilson "Bill" Nye; and "Whenever I feel the urge to exercise I lie down until it goes away," which in truth originated with the humorist J. P. McEvoy, author of the comic strip *Dixie Dugan,* who was only a teenager when Twain died.

But the quip about summer in San Francisco captures the cold reality of the city, which sits on a peninsula, surrounded on three sides by the Pacific Ocean and San Francisco Bay. Cold water upwells from the depths, surfacing along the coast, chilling the air. Inland, sunshine warms the ground, and the ground warms the air. The warm air over the land moves upward, sucking in the cooler air over the sea. The wind blows through funnels created by hills and skyscrapers. Fog forms. Summer days in San Francisco can be as cold as winter days, but fog and wind conspire to turn certain summer days into blustery baths of chilled mist that permeates the marrow.

None of this is to say that winter temperatures themselves cannot drop in San Francisco. Three months ago, in January, a winter cold snap resulted in temperatures below freezing, and just inland thermometers flirted with the twenty-degree mark. San Francisco's shelters filled with homeless people. Stephanie Schaaf, a spokeswoman for shelter operators, was quoted in the *San Francisco Chronicle:* "Yesterday we took in over 500 people, and usually our shelters are pretty full anyway, with maybe 275. We're breaking out mats, spreading them across the floors, in the hallways, pretty much anywhere there's open space." Governor Arnold Schwarzenegger declared a state of emergency and offered help for shelters and farmers. Twain, if he were still alive, might have had something to say.

Almost certainly, Twain also would have done a piece on the cryonics facility twenty-one miles from here, where humans, freshly dead, have been frozen and stored, awaiting a time when technological improvements will allow them to be thawed out, brought back to life, and cured of whatever killed them in the first place. "The coldest winter I ever spent," Twain might have said, "was the last fifty summers submersed in liquid nitrogen in a warehouse just outside San Francisco." It started when James Bedford arranged for his own preservation. In 1967, immediately after he was declared dead and in keeping with his wishes, artificial respiration and heart massage were administered. A three-man team pumped glycerol into his veins and cooled him with dry ice. Then they dropped him into liquid nitrogen.

As long ago as 1962, a professor in Japan froze the brain of a cat for 203 days, then revived it—if the brief presence of brain waves can be used as a measure of revival. A year later, an American physics professor wrote *The Prospect of Immortality*, proposing that freezing humans might not be all that unreasonable. The idea was simple: suspend the animation of the sick or dying, store them until the science of medicine develops techniques to treat their diseases, and then thaw them out and revive something more than just a few brain waves. After Bedford, advocates who supported freezing humans organized conferences and formed committees. In 1972, a manual was published. Commercial providers sprang up, offering what they called "cryonic services." Something like seventy patients—they are invariably called "patients"—are now stored in liquid nitrogen. Some patients have been stored virtually intact, while others, to save money, stipulated that only their severed heads be stored. One quickly grasps the short distance between a belief in the possibility of revival after immersion in liquid nitrogen and the belief in the possibility of inserting a newly warmed brain into the presumably still warm but empty-skulled body of a remarkably charitable donor. Advocates talk of "solid state hypothermia" and the use of surplus missile silos for storage. Advertised prices hover around a hundred and fifty thousand dollars, making immortality surprisingly affordable.

There is the Japanese cat, and various insects and frogs that freeze and thaw naturally with winter and spring, and the arctic ground squirrel, which hibernates at temperatures close to freezing, and a manual, and businesses with business licenses and investors and customers. All of this notwithstanding, one might not, a priori, hold out much hope for the likes of James Bedford. Despite the care taken in freezing him, at the temperature of liquid nitrogen — 321 degrees below zero — he would have suffered from the destruction of cellular membranes. Sharp ice crystals would have formed wherever glycerol failed to flush out the water. It seems unlikely that he will be successfully thawed. A reporter recently put the question of ethics to a cryonics professional. The reporter, writing in the first person, as if planning her own preservation, phrased the question somewhat delicately, along the lines of "What if you can't wake me up?" The response: "Well, you're dead. I don't see a problem with that."

Unable to sleep, I read Roald Amundsen's *The South Pole: An Account of the Norwegian Antarctic Expedition in the* Fram, *1910–1912.* Amundsen set sail in the *Fram,* a wooden schooner 128 feet long, with something like a hundred sled dogs on board. This was the very same *Fram* previously frozen into the northern ice by Fridtjof Nansen in 1893 and my second frozen caterpillar's namesake. Nansen rode the sea ice for three years before returning to Norway and eventually passing the ship on to Amundsen, ostensibly for a protracted scientific expedition to the north. On the way to the Arctic, Amundsen announced that the little ship would head north via a short detour to the South Pole — just a quick stop to beat Scott's expedition and allow Amundsen and his colleagues to be the first humans to stand on the earth's southernmost point. He was successful, in both beating Scott to the pole and surviving the journey. Nansen himself, a man whose stamina in the cold rivaled that of polar bears and arctic foxes, later referred to Amundsen as "almost superhuman."

At home, in my freezer, my pet caterpillars Bedford and Fram are frozen but presumably still alive, or in some sense alive, or at least potentially alive, at a temperature well below freezing but positively tropical compared to that of liquid nitrogen. Spring is coming, and it is almost time to thaw them out.

❄ ❄ ❄

In the cold of winter, pipes freeze, flooding houses. Ice dams form on roofs and back meltwater up under shingles. Bigger ice dams form on rivers, often under bridges, holding back the flow of water and causing flooding before finally collapsing to release torrents downstream. Boats are frozen into harbors. People slip and trip on the ice. The air, though it feels brisk and clean, picks up the carbon monoxide, nitrogen oxides, and other gaseous trash of cold-started cars, then holds it close to the ground in envelopes sealed by overlying warm air. Bumps and potholes form in roads. Cars take a beating from the combination of rough roads and thickened oil at a time when their metal and plastic parts are rendered brittle by low temperatures.

Well-built roads have cambers—rounded surfaces that shed water. When it is not wicked away by a good camber, water stands on the road surface, finds its way into cracks, and then freezes. The expanding ice pushes the road surface upward, creating unplanned speed bumps. When the ice melts, voids collapse, forming potholes, or the joints between two slabs of concrete separate, leaving a space between.

"It really has to do with water getting into the structure of roadways, freezing and thawing and that happening repeatedly," explained a Washington State official to a reporter who had asked about pothole formation. "That's why you see more potholes in the winter and spring than in the summer. It's that freeze-thaw process that breaks apart roadways." Seattle repaired more than 136,000 potholes in 2002, a bad year for roads. After a cold snap in Ohio, more than 200 claims were filed against the state seeking reimbursement for repairs to cars damaged when they ran over potholes. When asked why the government was so slow in paying these claims, a court spokeswoman said that people were not filling out their forms correctly. A Michigan drivers' association usually sees something like 2,000 claims a year, averaging six hundred dollars each, for pothole-related damage to cars. Near Detroit, more than 200,000 potholes are filled each year. "Potholes are kind of like geese," a Michigan spokesman told another reporter. "They don't come by themselves; they come in flocks."

Near Blacksburg, Virginia, the government maintains something it calls the Smart Road, used for pavement and transportation technology research. Among other things, the Smart Road has seventy-five snowmaking towers used to create nasty test conditions. In Alaska, the situation may be thought of more as one of trial and error tempered by budget shortfalls, with no need for snowmaking towers.

There is the famous Alcan Highway, stretching from British Columbia to the Alaskan interior and from there via different routes to Anchorage, Fairbanks, and Valdez. The Alcan is also known as the Alaska Highway, the Alaskan Highway, the Alaska-Canadian Highway, and the Road to Tokyo. It was a war baby built by the U.S. Army as a supply road during World War II. Before the road was finished, the Japanese occupied Attu and Kiska in the Aleutian Islands, far out in the Bering Sea, but nevertheless part of Alaska and therefore American territory. The Alcan route was designed to join airstrips used to deliver American airplanes to the Soviets, via

Alaska, in the lend-lease program that gave the Russians access to America's manufacturing capabilities. The route went from Dawson Creek in British Columbia through Whitehorse in the Yukon to Delta Junction, Alaska—just over fifteen hundred miles from a town few people could place on a map, through a town few people could place on a map, to a town few people could place on a map: a wavy line that ran northwest from somewhere north of Calgary to the middle of nowhere in Alaska. When it was finished in 1943, long stretches of the highway were thin gravel over bulldozed permafrost. In spring, the permafrost melted, turning the highway into a linear bog, a trap for vehicles of all descriptions. Logs were laid down in the manner of railroad ties, providing structure. The logs were eventually covered with gravel, and the gravel was eventually covered with asphalt. Over several decades, convoys of military vehicles were replaced by convoys of aluminum camping trailers pulled by jeeps, and later by motor homes. The stretches of road built over logs— the so-called corduroy road—were dug up and rebuilt with proper gravel foundations. Although the Alcan is open year-round, it is still a study in potholes, gravel breaks, poor shoulders, and unplanned speed bumps.

More interesting, more audacious, and more ridiculous than the Alcan is the Hickel Highway. To understand the Hickel Highway, one has to understand the mentality of resource development in Alaska. According to Walter Hickel, elected governor of Alaska by a margin of eighty votes in 1966 and later appointed U.S. secretary of the Interior, "On May 2, 1967, a DC-3 with half dozen of us on board flew through the mountains of the Brooks Range directly over Anaktuvuk Pass. As I looked over the long gradual ramp of the North Slope, where the continental divide slowly merges with the Arctic Ocean, a vision hit me, or call it intuition. I saw an ocean of oil. 'There's 40 billion barrels of oil down there,' I said."

To get to the oil, Hickel needed a road. In 1968, he had the state's

Department of Transportation bulldoze a road north to Prudhoe Bay, starting near Fairbanks. The plan involved little more than pointing a team of bulldozers toward the Beaufort Sea, putting them in gear, and hoping for the best as they blazed through spruce bogs, over mountain passes, and across tundra. "I drove the first six or seven miles myself," Hickel later said. "I got off and I told Jim, I said, don't you shut this thing off until you get to Prudhoe Bay." A month into it, progress was suspended when temperatures dropped to sixty-three degrees below zero. At fifty below, they started north again. The bulldozers scraped off the top layer of the tundra, cutting a smooth trail into the permafrost itself. In winter this left a fine road, but in summer the exposed permafrost did what it does, melting and then leaving behind a long, straight scar filled with water. Slowly, the ground immediately below the scar melted, and the scar deepened. The road was permanently closed a month after it opened in 1969, but large stretches remain visible today.

The cold is hard on machinery. At forty below, motor oil has a consistency close to that of warm tar. Pistons strain against it. Copper wire grows stiff and breaks. Windshields crack from the tap of road gravel or even from a bump in the road or uneven expansion. The stuff put on the roads to improve traction does not improve maintenance. Gravel put down for traction scratches paint and cracks windshields and headlights. Sand works its way between moving parts. Salt, which keeps water flowing at temperatures slightly below zero, eats into metal chassis and the bodies of cars.

It is not just automobiles that suffer. At Denver International Airport, sand spread on runways during a 2007 snowstorm became airborne in the wind and then scratched the windshields of fourteen planes sufficiently to lead to cracking. And then there is the space shuttle *Challenger*. Hot gas leaked around cold O-rings, causing an explosion of liquid hydrogen and liquid oxygen and killing seven astronauts seventy-three seconds after liftoff. The temperature on the launchpad that January morning in 1986 was thirty-six degrees, fifteen degrees colder than any previous launch and colder than the conditions under which the rocket's O-rings were tested.

There is, too, the issue of carbon monoxide. Carbon monoxide is colorless, odorless, and poisonous. It locks up the blood's hemoglobin, preventing it from carrying oxygen. Even low levels can be dangerous to people suffering from cardiovascular disease. Normally, the sun warms the ground, heating the air above it, and the warmed air rises. Carbon monoxide, nitrogen oxides, and other gaseous trash of civilization rise with it, to be scattered and lost in the wind. But during temperature inversions, air near the ground settles in, trapped by a warmer layer above it. Inversions are common in cities that sit in geographical bowls, surrounded by hills or mountains, such as Anchorage, Fairbanks, and Santiago. They are also common in cities that sit between the ocean and the mountains, where cold air blows ashore under warmer air, such as Los Angeles, London, and Shanghai. Cold engines working against thick oil pump out more carbon monoxide than they should, the result of incomplete combustion. It is this combination of temperature inversions and cold starts that gives Anchorage and Fairbanks a history of what the Environmental Protection Agency calls "non-attainment for carbon monoxide requirements."

Worse still is indoor carbon monoxide. Leaking woodstoves and heaters fill rooms with the stuff. On a cold day, in a tight cabin warmed by a leaky woodstove, carbon monoxide levels can quickly reach hundreds of parts per million. Expose yourself to two hundred parts per million for a few hours, and you will experience a headache, nausea, and extreme fatigue. Expose yourself to eight hundred parts per million, and watch for convulsions and unconsciousness. Death will come within three hours.

The cold floods houses. Water from cooking and bathing and even breathing condenses on cold glass, dripping through windowsills and into walls to leave water stains and encourage mildew. Heat leaks up through ceilings, melting the bottommost layer of snow on the roof. Beneath accumulated snow, water trickles down the roof and then refreezes when it hits overhanging eaves and gutters, forming ice dams. The pressure of expanding ice pries gutters away from the house, or the weight of the ice drags them downward. A gutter can take the fascia with it. Worse still, ice builds up along the bottom edge of the roof. Eventually, a pool of standing water is held against the shingles. The water freezes and thaws with temperature changes, working its way under the shingles and prying them up. Water leaking around roofing nails freezes, and the expanding ice pulls the nails from the wood. Eventually, water leaks through the roof itself, into the attic. Insulation is ruined, exacerbating the problem. Water pools on top of the drywall, leaking through, or finds its way into the walls, leading to mildew and rotten studs. Occasionally, homeowners armed with blowtorches, intending to melt the ice dams that are causing all the trouble, inadvertently burn down their houses.

And there are frozen pipes. In Anchorage, in a single winter, two fire hydrants were once pushed out of the ground by ice, frost-heaved to the point of breaking. Inside houses, expanding ice within pipes has nowhere to go but outward, through the wall of the pipe, first forming a bulge and then a crack. When the pipe cracks, the water flows. Frozen pipes flood something like a quarter of a million houses each year in the United States. Families return from vacations in Florida and Hawaii to find water streaming out through front doors and garages. Walk-out basements become aquariums with furniture floating in stairways. Specialized companies come to pump out the water, dry out what can be dried, rip out the rest, and renovate. If a frozen pipe is caught before it bursts, a thawing service can be called. For something like a hundred dollars, a thawing service uses electrical current to melt the ice. To avoid the charge, homeowners sometimes break out blowtorches to thaw pipes—the same blowtorches they might use on ice dams. The official advice from State

Farm Insurance: "Never try to thaw a pipe with a torch or other open flame. Water damage is preferable to burning down your house."

❄   ❄   ❄

It is April sixth and forty-five degrees in Anchorage. The road in front of my house is lumpy with melted ice, and my car bounces along as though I am on a rutted four-wheel-drive track in the Serengeti. In the low spots, thick pools of watery slush have accumulated. I head north, hoping to find cooler weather in the Talkeetna Mountains. The piles of snow stockpiled by plows through the winter have started to melt. As they melt, they metamorphose into piles of dark road grit, the whiteness of the snow running off as water and leaving behind the greasy detritus shed by an automobile culture and the grime from five months of plowing up gravel and sand. Other cars kick up a dirty mist that coats and recoats my windshield and headlights. It is not just the roads that are filthy. It is everything within ten feet on either side — cars, streetlights, sidewalks, fire hydrants, and mailboxes, all sprayed and resprayed by splashes of meltwater and, farther from the road, by meltwater mist.

Creeks, in flood stage, flow furiously beneath bridges.

Even at the long-abandoned Independence Mine, a gold mine sixty miles to the north and three thousand feet above Anchorage, the temperature hangs in the low forties, and the surface of the snow has turned to slush. On the steeper slopes, snow rendered unstable by the heat has collapsed, leaving avalanche scars of various sizes. In the summer, I have seen marmots here. Now they are under the snow, beginning to stir, hungry and suddenly eager for spring, with meltwater dripping into their rocky lairs and sunshine creating a blue twilight through the remaining snow.

Outside, I clip on cross-country skis. I force the skis through the slush around the basin that surrounds the mine. Skis are meant to glide, but as the snow melts, their motion changes. The first ski has to be pushed forward and the lagging ski dragged along. The snowshoes and waterproof boots that I left in my garage would be superior footwear in these conditions. I meander up to an old mining shack, abandoned on a hillside. The shack is built of sheet metal fastened to a frame of unmilled tree trunks and branches. Sod bricks ensheathe the ramshackle shell. Inside, the remains of a platform bed are surrounded by fallen pieces of sheet metal and an old white sink sitting on an earthen floor. Someone has taken the stove, which likely would have run on coal hauled up from seams that surface lower down the mountain. The miners' lives here would have been full of hardship and toil surrounded by scenic wonder. I sit on my pack in front of the shack, surveying the basin, feeling dreadfully tired, as depleted as the melting snow.

Gold was discovered early near Anchorage. By 1890, prospectors had spread from the first finds in the creeks south of Anchorage to the Matanuska and Susitna valleys to the north. In the creeks, they found color—"color" being slang for flakes of gold. They turned up the occasional nugget. They followed the creeks upstream, looking for the source of these flakes. As they moved closer to the source, the color grew richer, nuggets more common. The gold was being washed out of the mountains, chiseled out by expanding ice and then carried downward by meltwater into the streams and rivers of the valleys. The lucky prospectors were the ones who tempered their luck with knowledge, patience, toughness, and the backbreaking fortitude required to stoop over an icy stream with a gold pan that did triple duty as a cooking pot and a plate for dried beans and moose meat. The truly lucky prospectors found the sources of the gold that colored the creeks. They found veins of quartz with embedded gold, the mother lode. And there they set up hard-rock mines—tunnels into the solid rock itself.

Hard-rock mining was a different business than the placer mining of the creeks or the digging of tunnels through frozen soil to reach buried riverbeds. The word "placer" comes from the Spanish for sand bank, and placer mining was all about mining gold from sand and gravel—gold that had been washed away from veins in the mountains and mixed with common earth. In the area around Fairbanks and around the Yukon, placer miners used fires and steam to tunnel down through frozen silt and clay, reaching toward long-buried streambeds that had not seen flowing water since the Pleistocene. But here, in the Talkeetnas, the miners were boring into the rock itself, chasing the mother lode, pursuing the gold at its source. It was a matter not of melting permafrost but of dynamiting solid rock.

This sort of mining required capital and teamwork. Expertise was needed in explosives, in engineering, in geology. Miners who once worked alone or in teams of two or three formed companies. The Alaska Pacific Consolidated Mining Company eventually owned eighty-three claims that covered thirteen hundred acres. More than eleven miles of tunnels extended below the surface. On the surface, they built a town. By 1941, more than two hundred men worked in the mine. Twenty-two families lived in what amounted to a company town in the frozen mountains, many very difficult miles from Anchorage, which in the mine's early days was itself a muddy and isolated community. But the miners' children had their own schoolhouse, and the mountains gave up more than a ton of gold, worth more than twenty million dollars on today's market.

From the wide-open but often frigid outside, the miners rode trams into the tunnels. Underground, conditions were cramped. In winter, many of the miners would enter the shafts well before daylight and come out well after dusk. A day in the shafts would leave them wet with the sweat of hard labor and the moisture of the earth. Outside, at forty below, their clothes would freeze within seconds. This was a time of cotton and wool, a time before breathable synthetic fabrics.

When America entered the Second World War, gold mining was

deemed unimportant to the war effort. The government considered it nonessential. For a short time, the Independence Mine stayed open, claiming to produce scheelite, a source of tungsten needed for the war effort. But scheelite was a ruse that allowed gold production to continue. The authorities caught on and eventually shut the place down. The miners were sent home or to the battlefields of Europe or the heat of the Pacific. Some of them may have later sat in jungle clearings listening to lectures on the dangers of frostbite and hypothermia and the importance of layering clothing.

It starts to rain. In the flat light, the snow merges with the clouds, making it hard to know where the earth ends and the sky begins. Below, between me and the abandoned buildings of the Independence Mine, a diesel-powered snowcat—a tracked vehicle with a plow on the front and an enclosed cab for the driver—crawls up the trail, billowing smoke, working the snow. I ski down to it. The driver climbs down from the cab. The government pays him to groom the trails around the Independence Mine, attracting cross-country skiers. He tells me that he has had several flat tires this winter. The tires, which reside under the snowcat's tracks, pick up water from the snow. Water can settle in the bead of a tire, between the rubber and the metal hub, and when the water freezes, it expands and separates the tire from the rim. He claims that temperatures here at the mine did not go above minus forty for a month this winter.

I am strangely exhausted, as fatigued as if I had been huffing carbon monoxide. I am as tired as a miner just back from the tunnels, emerging from the darkness of a shaft into the flat light of spring.

❄   ❄   ❄

The *Anchorage Daily News* reports on Port Heiden, an Alutiiq village four hundred miles southwest of Anchorage. The sea has frozen for the first time in seven years, driving sea otters out of the bay. Starving, the otters slide and waddle along the ice and the frozen tundra. Dogs attack them, or they are killed for their hides, or foxes get them. One man reported seeing thirty-five of them sharing a small hole in the ice, taking turns diving for food, slowly starving. They shared space on the ice with half a dozen otter carcasses. While the otters foraged through the hole, eagles fed on their dead brothers lying on the surface.

Another story reported that the Canada geese had returned to Anchorage and then left. Twenty of them had been spotted flying over an urban neighborhood in late March. Another eight had touched down on a frozen lagoon two weeks later. Then, finding the city still more or less frozen in, they disappeared. "They may have gone back somewhere," suggested the director of Audubon Alaska. "They're not going to go all the way back to California, but they may turn around and go back just far enough to find what they need ... in Seward or Homer or maybe the Copper River Delta."

There is, too, the question of wood frogs and ground squirrels. Although wood frogs freeze solid, they will not survive temperatures below twenty-one degrees. If snow cover is light and air temperatures bitterly cold, their hibernaculums become graves. For ground squirrels, it is not so much a matter of freezing as it is of repeatedly warming and rewarming. It is a matter of letting their body temperature drop to the freezing point, shivering to warm up, and then cooling down again to conserve calories. Forced to warm up one too many times, they run out of stored calories and freeze solid. They starve and then freeze, like Arctic explorers.

Whenever low temperatures approach, gardeners talk of cold hardiness. The U.S. Department of Agriculture divides the country into zones based on minimum temperatures and illustrates these zones. Honolulu, an orangish red, stays above forty degrees. Miami, the color of faded bricks, does not drop below thirty-five. Fairbanks, the color of an unripened peach, can get colder than fifty below zero. In the orangish red zone, there are bougainvillea and royal palms and rubber plants. In the zone of unripened peach, there are cinquefoils and dwarf birches and scraggly black spruces that, even when perfectly healthy, look as though they have survived a fire.

Gardeners and farmers have tricks. They cover small beds of flowers with sheets to get through a night or two of frost. They choose their fields, avoiding low areas that trap cold air. During temperature inversions, wind machines and helicopters have been used to blow the cold air off the ground, away from crops. They cut channels through hedgerows to let cold air from fields drain into ravines. And they use smudge pots. Developed after a 1913 freeze wiped out a fruit crop in southern California, smudge pots burn oil to generate heat and sooty smoke. The smoke settles above the crop, blanketing the trees, trapping heat from the ground and from the pots themselves. Burning tires have the same effect. Misting with irrigation water has the double effect of forming a cloud that traps heat and extracting heat from the water itself as it turns to ice. But if the weather stays cold, too much of the mist turns to ice. Ice covers the plants, and the weight of the ice snaps off branches.

It is not just a matter of cold. Plants harden gradually to winter conditions, but when the weather breaks, they soften quickly. A sudden cold snap in autumn, before plants have hardened, is death. A temporary warm spell in late winter is death. A blanket of snow is a lifesaver. Leaves and stems are less sensitive than roots, so plants can tolerate cold air better than cold soil. In the end, a cold snap can wipe out a farmer as quickly as a drought. "There's nothing alive," a farmer in Illinois told an Associated Press reporter following an April 2007 freeze. "They're all dead. They say you pay your bills with apples and make your money with peaches. This year, you're not going to make anything on either side."

For humans, there is the issue of the cold itself, and winter damage to crops and roads and pipes, and fire hydrants frost-heaved out of the ground. But there is also the issue of slipping on ice. Sledders and skiers suffer various sprains and breaks and even death. States and cities have considered helmet laws for children on sleds.

Walkers, too, are affected. An oil field worker in northern Alaska slipped, apparently falling backward. The back of his head connected with the ice. He died.

And there are avalanches. A 2002 report from the Kenai National Wildlife Refuge describes the investigation of a herd of caribou caught in an avalanche the previous winter. The herd, wandering across a steep slope, triggered the slide. "The helicopter prop wash filled the sky with caribou hair," the report says, "and caribou skulls and bones lay scattered over a large area." At least 143 caribou died under thousands of tons of suddenly moving snow.

❄   ❄   ❄

It is April twenty-eighth here in Orlando, Florida, and pushing eighty degrees. On the North Slope, it is seven above. On the North Pole, the thermometer flirts with ten above. On the South Pole, temperatures are dropping below minus sixty.

In 1905 on this date, Orlando thermometers dipped briefly to sixteen degrees. In 1940, Miami dropped to twenty-eight degrees. In 1985, Jacksonville found the seven-degree mark, and Pensacola encountered five degrees. In 1899, Tallahassee experienced subzero temperatures, falling to two below. People in Florida hate the cold. Many have moved here from places such as Detroit and Minneapolis and Montana. My cabdriver, originally from Chicago, brags that he no longer owns a warm coat. Near the hotel pool, a woman who grew up in Florida tells me that she cannot even imagine forty below.

"Imagine thirty degrees," I tell her. "Now imagine eighty degrees. They are fifty degrees apart. Now imagine that same difference, but downward from thirty degrees. That is twenty below. Subtract another twenty degrees, and you have forty below." She stares blankly, as if at a madman. Imagination cannot extrapolate beyond the temperatures it has experienced. I stand in the sun, face upward, enjoying for the moment a latitudinal spring.

MAY

It is May fifth and hovering in the high forties on Prince William Sound, southeast of Anchorage, the site, seventeen years ago, of the 1989 *Exxon Valdez* oil spill. Prince William Sound is Alaska at its best: a piece of the ocean protected by snow-covered mountains, kelp and barnacles visible through clear water, puffins and otters on the surface, the occasional bear foraging along the shore, orcas and sea lions in the waves. Weather and time have washed away obvious remnants of the spill, but oil still sits beneath the gravel on certain beaches, and biologists continue to argue over evidence of residual damage. The temperature swings noticeably as our boat moves across the water, here finding a warm spot in the sun, here feeling the wind blowing down the slopes of a snow-covered mountain, here chilled by a glacier. Young cow parsnip sprouts velvety green along the shore, and willows bud on south-facing slopes. Spruce boughs stand dark green, having shed their winter burden of snow, but beneath the spruce, the snow remains deep. Higher up, slabs of snow have collapsed, dropping two or three feet straight down and leaving obvious faults in the snow. In places, the slabs have slid down the sides of mountains in avalanches of various dimensions. Where the mountains catch the sun, entire ridgelines have slid. Below, the avalanche snow lies in piles taller than trees, burying God knows what. The snow and ice have fractured pieces of the mountain's marrow, then pulverized them, leaving dirty streaks of mountain grindings exposed in fingers that reach downward across the snow-covered slopes.

On the water: flocks of kittiwake gulls, murrelets in groups of three and four, pigeon guillemots with their black-and-white wings bobbing in the waves, groups of Dall's porpoises as fast as torpedoes, scattered otters floating with their bellies and paws exposed to the sun, and three humpback whales. All but the humpbacks have overwintered here or nearby, tolerating the cold. The humpbacks are just back from Hawaii or Japan or Mexico, where they have fasted for months, focusing on singing their famous love songs under the waves, frolicking and courting and mating. Now they move patiently, submerging to sieve food from the rich, cold depths of Prince William Sound, then surfacing to breathe, then submerging to eat again. Their grazing is like that of elephants, the movements of slow-motion ballet carried by tons of fleshy momentum.

Our boat pushes in toward Beloit Glacier. Beloit is a tidewater glacier, reaching out past land into the bay itself, a grand remnant of a much bigger glacier that would have resided here in the late Pleistocene, shrinking and growing with warm spells and cold spells for untold years. Waves lap along the glacier's front, and ice thunders down directly into the sea, creating violent waves that radiate outward. Occasionally, ice calves off underwater and rockets upward, emerging like a frozen submarine. As we move closer, the boat's steel hull crashes through increasingly large blocks of floating ice. A mile from the glacier, we slow down and then drift with the ice, watching the glacier melt. Cold air sweeps down the glacier and out across a mile of water. The Windbreaker on a man standing next to me flaps like a flag, and his hood fills like a wind sock. The glacier's breath mocks us, reminding us that spring is delayed, that summer is short, that winter will be back soon enough. We live at the end of a time of glaciation and ice, in the warmish dusk of an ice age. The massive glaciations of the late Pleistocene were only yesterday.

❄ ❄ ❄

Spring came to the Pleistocene Ice Age about ten thousand years ago. Prior to the start of this recent interglacial, two great ice sheets and many smaller ice fields and glaciers stretched across North America during a period of glaciation that lasted a hundred thousand years. Today's Beloit Glacier in Prince William Sound is to those ice sheets as a lemming is to a woolly mammoth. During the periods of extensive glaciation, the sites of what would become Chicago, Seattle, Boston, Cleveland, and Kansas City were perpetually snowed in. Just to the south, there was tundra, and to the south of that, forests and grasslands that were stocked with horses and small camels and mastodons. Ground sloths the size of elephants lumbered about, hiding from saber-toothed tigers and American lions. The short-faced bear stood more than five feet tall at its shoulders. Reared up on its hind legs, it stood twelve feet tall. It has been said that the presence of this bear slowed the movement of humans across the Bering Land Bridge, a now drowned stretch of land also called Beringia.

Even during the coldest periods of the Pleistocene Ice Age, some areas were surprisingly ice-free. Mountaintops stood above the ice. Along the east and west coasts of North America, certain mountains blocked the moving ice sheets, leaving shadows of open but sparsely vegetated ground. Strangest of all, most of interior Alaska, too dry to generate massive quantities of snow, was free of the glaciers that blanketed most of northern North America. Alaska was windswept and brutally cold, but not entirely buried in mile-thick snow and ice. And Alaska did not look like Alaska. With so much of the world's water tied up in ice, sea level was three hundred feet lower than it is today, and the Bering Land Bridge, as ice-free as the rest of the state, joined North America and Russia. What is now St. Paul Island, a lonely postal address in the middle of the Bering Sea, was then a low hill in what was likely a grassy plain. But the grassy plain did not look like today's grassy plains. Instead it was a mix of Arctic tundra and modern grassland, sometimes called the Arctic steppe. In its abundance of wild animals and in their variety, the Arctic steppe resembled the Serengeti.

Wildlife in Beringia, on the Arctic steppe, was not the same as that found south of the great ice sheets, but it was similar. Mammoths wandered in the grass with musk oxen, bison, elk, grizzly bears, and Dall sheep. During the warm interglacial periods that came and went during the two and a half million years of the Pleistocene Ice Age, the corridors joining north to south would open like icy gates, letting animals through from each direction. But if the warmth stayed too long, sea level would rise and flood Beringia, closing the watery gate between Asia and America. The Beringia gate closed each time the sea rose to within 125 feet of its current level.

The camel and the horse, both products of North American evolution, found their way north and eventually into Asia while the gate was open, and later went extinct in North America. The gray wolf found its way east from Europe and then south, where it met its much bigger and now very much extinct distant cousin the dire wolf. Deer and sheep may have arisen in the old world and crossed to the new via Beringia. Each time the gates closed, the animals locked in on either side would evolve on their own.

The ice gates — the massive glaciers and snowfields — separating north from south opened and closed to some degree in concert with the Beringia gates. The animals south of the ice evolved differently than those north of the ice, developing esoteric differences. The bison of Beringia had two shoulder humps and long curved horns, while the bison south of the ice had a single hump and shorter horns. At one time, the grizzly bear may have been found only north of the ice, while the black bear may have been found only to the south. The Dall sheep was found to the north, while the bighorn sheep was found to the south.

The ice created its own weather. Wind whipped off the edges of the ice sheets. Where ice sheets met the ocean, they chilled the water, powering currents. Warm air meeting the ice cooled, and its moisture fell out as snow. Air blowing into the interior of the sheets tended to be sucked dry. Over time, the interior ice sheets thinned, snow-starved. During the Pleistocene Ice Age — the most recent ice age, the one that is dying now, choked by greenhouse gases — spring came and went repeatedly. Each time that metaphorical spring came to the earth, the ice would retreat, biding its time through a millennial summer and then advancing again. And each spring, in retreating, the ice would dump its water into the sea. Sea level would rise reasonably quickly.

The land rose, too. Relieved of the weight of ice, the land, for ages pressed down by the sheer mass of the ice it had carried on its shoulders, would rebound. Areas would flood with the rising sea level and then emerge, springing upward with the burden of ice removed. Dry pastures and hilltops and mountainsides would emerge with fossils of shells, fish, crabs, and in at least one case kelp, all speaking of a time when they were flooded by cold water.

The Canadian ecologist Evelyn Pielou put the timing of the Pleistocene Ice Age in perspective. "To make the relative lengths of enormous stretches of time easy to visualize," she wrote, "let us use as a model one decade to represent the past billion years." That would make the earth about forty-five years old. "On the scale of the model," Pielou continued, "a glacial age lasts a month or two." One great ice age occurred seven or eight years ago, and another just two or three years ago, when the continents shifted into position to separate the steady warmth of the tropics from the seasonal cold of the far north. Each lasted a month or two, until the continents could drift a bit, opening a gate through which warm currents could carry warm water north and cold currents could dump cold water south. The Pleistocene started last week. Things seem warm now, with the great ice sheets pulled back and no more than a patch of snow covering Greenland, another over the North Pole, another to the far south, and a few scattered snowfields and glaciers here and there. During this last week, the ice has expanded and pulled back and expanded and pulled back again and again, at least nineteen times, responding to the level of carbon dioxide in the atmosphere, the liveliness of the sun, and the Milankovitch cycle of an always changing tilt in the earth's axis and a stretching out of its orbit. If greenhouse gases do not kick this cycle out of kilter, as they almost certainly will, the next expansion of the ice is only minutes away. Minutes being, of course, thousands of years.

An hour ago, spring came to the Pleistocene. At first, the ice sheets thinned without pulling back. The thinning ice dumped water into the oceans. Sea level rose. The shoreline of northern North America was for the most part one long stretch of tidewater glacier. But then the ice sheets pulled back, melting at their edges. The earth's surface, relieved of all this weight, rebounded. Bay bottoms and coastal waters became mudflats and then salt marshes and then forests. What is now Lake Champlain between New York and Vermont was then part of the Champlain Sea, which covered Ottawa, Montreal, and Quebec City for two thousand years—actual years, not the fast-forwarded-model years needed to put vast lengths of time into perspective, but two thousand winters and summers, twenty centuries. Its early shorelines were cliffs of ice. Bowhead, finback, and humpback whales swam between those shorelines. Harbor porpoises frolicked about. Ringed seals basked on the frozen surface in early spring. As those two thousand years went on, the land rebounded. The Champlain Sea drained, leaving little more than Lake Champlain, the fossils of whales and seals and cold-water clams, and isolated patches of beach grass and sea rocket far inland from the Atlantic. The same pattern happened all along the coast. In Maine, Augusta and Bangor rose up from underwater. The now extinct Tyrrell Sea shrank to become Hudson Bay, surrounded by rows of terrestrial beach terraces that speak of a former glory.

On land, retreating ice left barren ground, scraped of all vegetation, covered with rock rubble and piles of boulders and stones ground into flour. Wind sliding down the face of the ice sheets tossed the flour into violent sandstorms. In places, great blocks of ice fell from the faces of retreating ice sheets. Blowing dirt built up around them, and when they melted, they left in place the kettle lakes of the prairie states and Washington and New York states. In other places, the blowing dirt—sand and flour that the glaciers ground from bedrock—covered vast areas of ice, insulating it. This ice, insulated beneath soil, stagnated. Grasses and later forests grew in the soil, with the stagnant ice beneath. Eventually, the stagnant ice melted, and the ground, deprived of the subterranean ice, subsided. What had been upland forests sank into lowlands, wetlands, and swampy depressions.

One seed at a time, plants moved onto the quickly changing ground. Marsh marigold, mountain monkshood, and mountain harebell all moved south from Beringia. Plants with wind-borne seeds moved faster than those that hitched rides in the guts of animals. Hickories were among the slowest, taking two thousand years to journey up the Mississippi Valley and then east to Connecticut. Chestnuts, too, crawled along, averaging a mile or so every ten years. Hemlocks and maples were twice as fast. Certain oaks sprinted at more than two miles every ten years, neck and neck with eastern white pines.

As the land changed and new plants arrived, the forests changed. Pines replaced spruces. Balsam firs, birches, elms, and oaks replaced pines. At certain places and certain times, forests changed during the course of a human lifetime.

As the ice melted and the land rebounded, great rivers and their drainages changed, too. Arctic grayling, northern pike, and lake whitefish survived glaciation in the Yukon River and its tributaries, then migrated into the Mackenzie River. They did not migrate along coastal waters, but rather through a lake that formed as the glaciers melted, a lake that joined the two rivers, another open gate. Great Bear Lake, Great Slave Lake, and Lake Athabasca were at one time part of Lake McConnell, which drained for a time into the Mississippi River and for a time into the Mackenzie River and then into what would become Lake Superior, switching back and forth as floodgates of ice opened and closed. Certain fish evolved for a while behind the gates, then spread out when the gates were opened. The northern pike and burbot of Alaska hail from Beringia, but the northern pike and burbot of the lower states hail from waters south of the ice sheets. Genetically, they are not the same.

Humans passed through the gates more than once. They came, it seems, as early as forty thousand years ago, but these early North Americans left little behind. What little they did leave was primitive and crude, nothing more than rocks with an edge, the kind of thing that one might pick up and skim across a lake without noticing that it had once been worked by human hands. The more recent Clovis people left stuff lying around, pretty stuff, symmetrically carved functional art. The first remnants of the Clovis people found in modern times were in New Mexico, but ongoing searches have turned up Clovis sites throughout the United States and down into Mexico and even Central America. At sites dated to thirteen thousand years ago, the Clovis people left spear tips and knives and scrapers made of chipped stone, bones marked by the signs of butchering, and rocks burned red. Most abundantly, they left stone flakes, fine chips of stone, the waste from working rocks into more useful tools. They ate more than 125 species of plants and animals, including mammoths. Although what they wore is unknown, it seems likely that they would have had the sense to wear skins in cold weather. Closer to the ice, with the wind ripping down from the ice sheets, it seems obvious that they would have worn one layer of fur turned in toward the body and another with the fur turned out.

Around the time of the Clovis people, the mammoth disappeared. Other animals believed to be on the Clovis people's menu also disappeared: North American camels, two kinds of deer, two kinds of pronghorn, a kind of llama, the stag moose, the shrub ox, the woodland musk ox, five species of the American horse, and mastodons. Giant beavers, as big as today's black bears, disappeared. The American cheetah, the dire wolf, the saber-toothed tiger, the short-faced bear, and the American lion all disappeared. Smaller mammals — the sort some believe less likely to show up on a menu — survived. Certain scientists saw this as circumstantial evidence and blamed the Clovis people for the extinctions, but other scientists argued that there were more clues to consider. To anyone who has hunted an elephant or a musk ox or a moose with a stone spear, the starring role of man in these extinctions seems miscast. More likely, humans helped the extinctions along as the animals succumbed to a rapidly warming environment — roaring winds blowing from melting glaciers and sandstorms carrying megatons of glacial flour. They succumbed to a sudden change that likely made winters and summers even more uncertain than they would be during the Little Ice Age. It was a time when grasslands became forests, forests were buried in drifting sand, and massive lakes drained overnight — a lake one day, a mud bed the next — through white-water torrents bigger than today's Mississippi River. Tallgrass prairies became shortgrass prairies. The Arctic steppe of Beringia became overgrown with shrubs and then trees before disappearing altogether under the rising Bering Sea. Disease, too, may have played a role in these extinctions. But this much is certain: the big mammals were there, south of the ice, and they were in Beringia, north of the ice, and when the ice melted, many of them disappeared.

Within a few hundred years of the extinctions, on the other side of the world, the Sumerians were busily inventing agriculture. But in North America, the hunt went on. Bison were still abundant, and various deer, and the delicately sized and flavored modern beaver. Grizzly bears kept the hunt exciting. Clovis boys sat around campfires and, with the timeless cockiness of teenagers, mocked the stories of ice sheets and mammoths and mastodons, and of a great-uncle who had been eaten by a saber-toothed tiger. The boys focused on their own prowess at hunting buffalo and on near misses with grizzlies. But occasionally they stumbled upon bones, upon tusks and teeth of mammoths and mastodons, and they may have wandered through boulder-strewn forests and grasslands, perhaps even having their own word for erratics, wondering why such large rocks would be resting so far from anything resembling a mountain.

❄ ❄ ❄

It is May sixth and warm in Anchorage, truly spring. To celebrate, I take my caterpillars Fram and Bedford from the freezer. They have been on ice since September twenty-third. I put the frozen but presumably undead bodies of my two patients in a mason jar lined with the budding leaves of birch and willow and sambucus. Optimistically, I poke airholes in the jar's lid. I also take out my frozen mud, collected in September and stored in the freezer ever since. I open the jar to let the mud thaw.

The Anchorage paper runs a full-page article on mosquito evolution. For the past five years, a pair of scientists have created the climate of New Jersey in an Oregon laboratory. The climate chambers have been stocked with mosquitoes from Maine. From the perspective of scientific inquiry, storing Maine mosquitoes in Oregon climate chambers that mimic New Jersey conditions is business as usual. The way things are going, Maine's climate will be New Jersey's climate in the foreseeable future. The mosquitoes have already shaved two weeks off their hibernation time. "In a woodsy bog on the road between Millinocket and Baxter State Park," the paper says, "a mosquito that can barely fly is emerging as one of climate change's early winners." The mosquito may have had a trick or two to show the mammoth, North American camel, llama, deer, pronghorn, stag moose, shrub ox, woodland musk ox, American horses, and mastodons.

❄   ❄   ❄

Mammoths occasionally materialize out of thawing soil where rivers cut into banks of permafrost or where miners dig into icy gravel. Their tusks stand out, or blackened femurs as big as fence posts, or skulls with a large central aperture for the trunk. The skulls at different times have been mistaken for the skulls of unicorns and Cyclopes. In the past few centuries, more than fifty thousand tusks have been exported from the Taymyr Peninsula in Russia. Today Alaskans market mammoth tusks as expensive souvenirs.

In 1977, a Russian miner working near the Dima River found a frozen carcass—not a skeleton, but a frozen carcass, covered with hair, frozen eyes intact and staring. In his excitement, he reportedly called out, *"Mamonyonok!"*—"Baby mammoth!" Frozen mammoths tend to be named. This one was named Dima, after the river. Before the miners could attract the attention of officials, Dima's carcass thawed enough to stink. Nevertheless, Dima's heart and foot-long penis are displayed today in the St. Petersburg Zoological Museum.

Dima was neither the first nor the last. Chinese writings from the second century B.C. describe thawing mammoth remains, saying they were "found beneath the ice, in the midst of the ground." The Chinese text talks of flesh weighing a thousand pounds that "may be used as dried meat for food." It indicates that rats flocked to a thawing carcass: "Wherever its hair may be found, rats are sure to flock together."

In 1901, the Russian Imperial Academy of Science heard of a

carcass frozen in a cliff on the Berezovka River, above the Arctic Circle. Scientists traveled by Siberian Express, wagons, boats, and horse-drawn sleds to reach the site. By the time they arrived, wolves and foxes had devoured part of the mammoth. Local people had taken the tusks. Eugene Pfizenmayer, one of the scientists, wrote:

> Some time before the mammoth body came in view I smelt its anything but pleasant odor—like the smell of a badly kept stable heavily blended with that of offal. Then, round a bend in the path, the towering skull appeared, and we stood at the grave of the diluvial monster. The body and limbs still stuck partially in the masses of earth along with which the corpse had been precipitated in a big fall from the bank of ice.

The scientists built a shelter over the mammoth and went about dismembering what was left of it. They learned that it had four toes and a flap of skin protecting its anus from the cold. They preserved its flesh with alum and salt, then shipped the whole thing to St. Petersburg. One of them wrote, "A thorough washing failed to remove the horrible smell from our hands." The trip back to St. Petersburg, via sled at temperatures as low as sixty-seven below, began on October 15, 1901, and ended on February 18, 1902. Summed up, the expedition north, the butchering of the frozen mammoth, and the trip home took 291 days. Tsar Nicholas and Empress Alexandra viewed the carcass in St. Petersburg. The empress held a handkerchief to her nose. "Is there something else interesting to show me in this museum," she said, "as far away from this as possible?"

In 1999, a frozen cube of permafrost with a mammoth inside was excavated from the Russian tundra. They named this one Jarkov, after the father of the boy who found the carcass. By the time scientists showed up, the tusks had been sawed off and sold. Excavation trenches filled with meltwater. Fuel shortages delayed the arrival of jackhammers. Eventually, the cube of permafrost containing Jarkov was fitted with a metal lifting frame. For the sake of television, new tusks were bolted to the frozen cube. The twenty-three-ton block of permafrost was coaxed into the air by a helicopter rated for twenty tons. The metal frame bent, but the mammoth flew, bolted tusks grandly protruding from the ice. It landed in its new home—a temperature-controlled room, a glorified ice cellar dug into the permafrost. French scientists slowly thawed parts of the cube with a hair dryer. They found Jarkov's flesh and hair. They also found bones out of place, suggesting that the mammoth was less than perfectly intact. Perhaps in part for the sake of drama, the scientists, in matching overalls, stopped short of disinterring Jarkov. The mammoth remains exhumed but frozen, preserved until the right scientists with the right amount of funding and the right questions come along to thaw it out. Jarkov lies in state—like Lenin, but perhaps colder and less famous.

Frozen mammoths carry with them frozen stomach contents. They have buttercups frozen between their teeth. They may have frozen clues to their extinction, too. If a plague killed the mammoth, saber-toothed tiger, and North American camel, freeze-dried evidence may persist. Certain scientists build careers around looking for the kind of species-jumping virus that might be responsible for the Pleistocene extinctions. They look for something like the AIDS virus, which jumped from monkeys to humans; elephant herpes, which jumps between Asian and African elephants; or rinderpest, which jumps between buffalo, wildebeests, hartebeests, and bongos, causing fever, infection around the mouth, tissue necrosis, and death.

In some cases, the people living among the frozen remains can claim ancestors who lived among the unfrozen reality of living mammoths. These people often believe that the frozen flesh is dangerous, that it brings fatally bad luck. At one time, certain Siberian natives believed that the creatures were large rats that lived underground but died when exposed to the sun. Others believed that they lived in the mountains but came down to feed on human corpses. Until Victorian times, European scientists argued that the remains were nothing more than African elephants swept to the Arctic by the biblical Flood. Thomas Jefferson, believing stories from Native American tribes in the West, suggested that mammoths might still survive in the American interior. Jefferson tasked Lewis and Clark with confirming these stories.

In 1796, based on his comparison of mammoth bones with those of existing elephants, Georges Cuvier suggested that the mammoth might in fact be extinct, gone forever. The mammoth, he believed, was adapted to cold climates. The African elephant and the Asian elephant were not. Cuvier was among the first to articulate the possibility of extinction. "All of these facts," he wrote, "consistent among themselves, and not opposed by any report, seem to me to prove the existence of a world previous to ours, destroyed by some kind of catastrophe." In 1887, the paleontologist William Berryman Scott, who somehow mistakenly believed that the mammoth was carnivorous, considered the extinction a blessing: "The world is a much pleasanter place without them, and we can heartily thank heaven that the whole generation is extinct."

Not everyone would agree. Certain scientists, serious men who publish articles in prestigious scholarly journals, hope to clone a mammoth. The plan is to use DNA salvaged from frozen flesh. One vision involves removing a viable nucleus from a frozen mammoth and inserting it into a single egg cell from an elephant. The fertilized egg would be implanted in an elephant. Some two years later, if all went well, a baby mammoth would appear, brought back to life — an extreme survivor of hypothermia, superior to a seasonally thawed caterpillar, superior to the frozen human sperm that have been thawed and put to work after only a few years below zero, superior to James Bedford, even assuming that he is ever successfully resurrected from his bath of liquid nitrogen.

But to date, viable cells and viable researchers have not connected.

Instead, the meat has gone to museums or foxes or wolves or dogs. Or it has rotted on the tundra, or been tasted by field scientists. "It was awful," said one man who tasted a specimen believed to have died more than twenty thousand years ago. "It tasted like meat left too long in a freezer."

❄ ❄ ❄

It is May sixteenth and eighteen degrees on a man-made island in the Beaufort Sea. The island is six acres of gravel piled on the seabed, surrounded by steel sheet pilings and concrete blocks and cramped with oil wells, heavy equipment, and metal buildings full of pipes and tanks and gauges. To the north, six-foot-thick ice stretches to the horizon. To the south, ice stretches to the shore and then gives way to snow and the industrial complex of Prudhoe Bay. An ice road stretches across the sea ice from the shore. Sun glares off the ice.

I stand around watching a dive crew work. With an excavator, they have cut a moat in the ice to reach concrete blocks that form the island's first line of defense. In winter, the ice wreaks havoc with the concrete blocks. In summer, when the ice pulls away from the island, wind rips off the pack ice farther north and kicks up waves that slam into the island, wreaking more havoc. The divers will replace the blocks. For now, as a preliminary step, one of the divers is clearing ice that the excavator missed. He wears a yellow helmet attached to an umbilical that comes to the surface. The umbilical includes a hose that sends air to the diver's helmet and another hose that sends hot water to the diver's suit. The diver stands on the bottom, surrounded by ice water but soaked by a never-ending hot shower from the surface. The diver blasts away with a fire hose connected to a boiler on the surface, pumping hot water against ice-coated concrete blocks five feet below. A typical dive runs several hours.

One of the divers tells me that a polar bear wandered in last year, forcing them to end a dive on short notice. On the surface of the moat, piles of slush and blocks of ice bump the diver's umbilical. When the sun drops low in the sky, the temperature drops with it, and a thin film of new ice forms.

Despite all of this ice, it is spring. The sun, though it dips close to the horizon, will not set for another two months. Seals bask on the ice like plump tourists on a white sand beach. A pair of ravens guard a nest built high up in a pipe rack. Melting snow drips from the roofs of buildings. I talk to a worker who does two-week stints on the island every month. These six acres of gravel and machinery and oil wells are his home twenty-five weeks each year, half the year minus two weeks for an extended vacation. This is his sixth year. "I love the sound of dripping water," he tells me, watching meltwater trickling down from metal roofing.

❄   ❄   ❄

As the world warms, more remains pop out of the permafrost. With some regularity, mammoth hunters in Siberia stumble on human carcasses. Some are said to be victims of the Stalin years. Others may have died during smallpox epidemics. Permafrost is no place to dig a grave. Shallow graves would be the norm for those in a hurry. Stalin's executioners would not have made time to chisel down into the permafrost. Survivors disposing of plague-ridden corpses would have been equally harried, eager to be done with the dead and impatient with the hard ground.

In 1845, the entire Franklin expedition, 129 men, vanished in the snow and ice of the American Arctic. The disappearance prompted years of search parties, serving the dual purpose of looking for Sir John Franklin and looking for new scientific and geographic information. The search parties added scattered frozen corpses of their own to those of Franklin and his men. For several years after Franklin's disappearance, there was real hope of finding survivors. Elisha Kent Kane's expedition, mounted in part to find Franklin, survived three years in the Arctic. Likewise, a few members of the later Greely expedition survived three years. In Franklin's case, hope was misguided. Franklin and all of his men died north of the Arctic Circle. Some became food for the others. In 1984, a team of anthropologists disinterred a few of the bodies. These were the early victims, who were buried properly, in marked graves on Beechey Island, before the expedition fell apart and the remaining men understood how desperate the situation would become. The graves of these early victims had been found in 1850, during the search for Franklin and his crew, when hope remained that Franklin and some of his men still might be found alive. The anthropologists opened the graves to find the skin and clothes of the properly buried men intact, preserved by ice. Their hair was intact. Their eyes were frozen open. Their lips were pulled back, exposing bad teeth and signs of scurvy.

The Iceman of the Alps, found in 1991, had been frozen in the Tyrolean mountains for five thousand years. Two hikers spotted the carcass. At first it was thought to be just another dead climber. At least five or six dead climbers had already surfaced that year, some after spending as much as fifty years in the ice. But as the Iceman's body was hacked from the ice with picks and a ski pole and a jackhammer, it became apparent that he was somewhat more ancient than originally expected. It became apparent that he was a frozen chunk of the Neolithic. He was five feet two inches tall and in his late forties. When found, he was freeze-dried, drained of the moisture of life, and weighed 30 pounds. In life, he would have weighed something like 140 pounds. He had an arrow wound in his left shoulder. His arteries were clogged. His last meal had been ibex and red deer meat with bread baked from an early form of wheat. He wore a fur hat, a cape of woven grass, a fur wrap, leather leggings, and leather boots. The boots were insulated with grass. He carried a longbow, a stone knife, and a birch-bark cylinder that probably held live embers for starting a fire.

More recently, in 2005, the frozen body of Leo Mustonen was

found at the bottom of a glacier in the Sierra Nevada. He had been missing for sixty-three years, along with three others who never returned to Mather Field in Sacramento after a training flight. Leo wore a parachute with the word ARMY stenciled across it. He wore a torn sweater and a badly corroded metal name tag. He carried a comb, fifty-one cents in change, a Sheaffer pen, and an address book. An otherwise illegible note in his address book said, "All the girls know."

Franklin's disinterred men — the only complete remains known from the Franklin expedition — were reburied in place on Beechey Island, overlooking the sea. The Iceman lies frozen in an Italian museum. Leo Mustonen was cremated and his ashes were buried next to his departed mother in Minnesota. Leo's niece was quoted as saying that it was "nice to know he won't be left alone up in the mountains in a pile of snow."

❄   ❄   ❄

It is May nineteenth and about fifty degrees in the shadow of Eklutna Glacier in Alaska's Chugach State Park. My companion and I bicycled in on a rough gravel road for twelve miles. The gravel road was lined with birch and aspen and poplar in their full spring foliage, no longer simply budding but truly leafed out. At the end of the gravel road, we abandoned the bicycles and walked another mile through patches of snow and across shattered rocks and boulders as big as cars. We scampered over bedrock carved out by ice and smoothed by meltwater. Throughout, our dog trotted along behind, but here, just below the glacier, we surprise a mountain goat. The goat lures our dog along, staying just ahead of him, moving up toward the glacier. Both dog and goat ignore our calls.

Where we stop, waiting for our dog, the glacier has resided recently, probably within our lifetimes. A few wind-borne seeds have taken hold. Dryas, not yet in bloom, lines cracks in the rocks. Here and there, saxifrage has gained a toehold in dirt-filled depressions — its lavender flowers are open — but most of the ground is bare rock. A steady wind blows off the glacier and across the rock. Up the valley, what is left of the glacier is profiled with steep ridges and crevasses of white snow and blue ice. It looks distinctly like a glacier on its way out. Just below us, meltwater paints a pond azure. A quagmire of wet glacial flour surrounds the pond. Farther down, where glaciation is a somewhat more distant memory, patches of shrubs and even trees grow along the slopes. In places, the ground has collapsed, liquefied by melting snow and failing under its own weight or carried away in avalanches. Along the edges of the collapses, the young soil stands exposed, a fragile veneer, a thin skin creeping in behind the glacier. Far down the valley, past the collapses, the landscape is similar to that of Scotland, scarred by long-gone ice but more or less healed.

The goat reappears a quarter of a mile up the valley. Our dog comes limping back. The pads of his feet have been torn open, and he trembles in pain and fear. Out of sight, something happened. The goat butted him, or he tripped and slid on rocks and ice.

This is the sort of scene that would have been common when the Pleistocene ice pulled back: bare rock, patches of snow, struggling vegetation, constant wind, unstable soil, and animals that do not know how to behave.

JUNE

It is June second and just over sixty degrees in London. My taxi driver talks of plans for an afternoon on a Brighton beach after he has dropped me at Heathrow Airport. He insists on pointing out the sights. When he points out Kew Gardens, I ask if he knows that a polar bear skull was found there. During the Pleistocene, I tell him, polar bears roamed through what would become downtown London. "The bears," I say, "were as big and white as German and American tourists visiting Westminster Abbey." We ride the remaining twenty minutes in silence.

I fly for six hours above seas recently thawed and land that was glaciated only yesterday, and then I sit in Boston traffic. The thermometer stands north of eighty degrees. Cars battle for position, each belching carbon dioxide at an ice-age-killing rate of something like eight tons per year. In my midsize rental, I turn on the air conditioner. Prior to this, I have not turned on a car air conditioner in at least ten months. The temperature drops abruptly. As abruptly, my clothes turn clammy. King James would have felt this same sudden clamminess upon entering Cornelis Drebbel's air-conditioned Westminster Abbey four centuries ago. For the king, a sweaty king unaccustomed to air-conditioning and bearing the weight of royal clothes, this feeling of clamminess must have been overwhelming.

I stop at an old house south of Boston, a Cape built for cold weather. A sign outside says MARITIME MUSEUM. Inside, it is as much a museum of shipwrecks as a museum of ships. The curator, an elderly woman full of energy and enthusiasm about ships and shipping, has never heard of Frederic Tudor. She knows nothing of the Boston ice trade to the Caribbean and India. The Ice King means nothing to her. She seems hesitant to believe that ice was commercially harvested from Walden Pond and sold in the tropics.

She shows me a sketch depicting a wreck on a sandbar. Four men are perched atop the splintered remains of a mast. A fifth man is in a life ring of sorts, riding a cable stretched to shore. "They save the youngest first," the old woman tells me. "They have longer to live." The sketch, even without color, captures blowing snow and sleet. The hair of one of the men clinging to the mast is blown forward, hiding his face. In conditions like these, their clothes would have been stiff against their skin. Since they were wet and poorly protected, their core temperature would have quickly dropped, their hands numbing and their grip on the mast loosening, their will to live diminishing.

Outside, I am struck by the reality of Boston: traffic and heat and people in cars as big as fishing dories, all doing their level best to pump the atmosphere full of carbon dioxide, all doing their part to warm the planet. They are killing what little is left of their ice age. I stand by my own car for a moment thinking about my plane ride across the Atlantic and knowing that my one seat was responsible for something like half a ton of carbon dioxide. I've already dumped another ten pounds from my rental car's tailpipe in Boston.

"It's my ice age," I say to myself, "and I'm killing it."

❇   ❇   ❇

The Frenchman Joseph Fourier, orphaned at the age of eight, was active during the French Revolution. As a result, he was awarded an appointment at the École Normale Supérieure and eventually a chair at France's prestigious engineering school, the École Polytechnique. He probably knew something about the volcanic eruptions that caused the Year Without Summer. He served under Napoleon and would have heard of the devastations of frostbite suffered by the French army in Russia. He would have heard of soldiers burning themselves while rewarming frozen digits and limbs over open fires. He would have known something of the death of Vitus Bering and of early British attempts to navigate the Northwest Passage. He likely knew that ammonia could be liquefied at temperatures well below freezing. He may have read of early refrigerators.

In 1827, Fourier published an essay in which he recognized that certain gases in the atmosphere contributed to the warmth of the earth. Carbon dioxide, water vapor, and methane blanketed the earth, allowing visible light to pass through but absorbing warmth that was reflected back up. This was early in the Industrial Revolution, and these were for the most part naturally occurring gases. What Fourier described would later be called the greenhouse effect. Without the greenhouse effect, the average temperature of Miami might approximate that of Bangor, Maine. New York City would be as cold as Fairbanks, Alaska. Barrow, Alaska, would be substantially less inhabitable. On the whole, the earth without the greenhouse effect would be only marginally more tropical than Mars.

Fourier harbored a strong aversion to cold. He believed that wrapping up in blankets would improve his health. In 1830, wrapped in blankets, he tripped down a flight of steps. The fall killed him.

The Swede Svante Arrhenius was not born until 1859, almost a decade after the end of the Little Ice Age. By then carbon dioxide had been frozen, Agassiz had become famous for his belief in the great Pleistocene Ice Age, and Lord Kelvin had developed a temperature scale with zero set at the limit of molecular motion. Arrhenius probably rode, or at least saw, an early bicycle, a direct descendant of the Draisine, invented when the Year Without Summer pushed the cost of grain and hay beyond reach and rendered horses unaffordable. In the April 1896 edition of the *London, Edinburgh, and Dublin Philosophical Magazine and Journal of Science,* Arrhenius wrote of the greenhouse effect. "Is the mean temperature of the ground in any way influenced by the presence of heat-absorbing gases in the atmosphere?" he asked. "Fourier maintained that the atmosphere acts like the glass of a hothouse, because it lets through the light rays of the sun but retains the dark rays from the ground." Arrhenius went on to estimate that a doubling of carbon dioxide in the air would lead to a five-degree increase in average temperature. He believed that natural increases in carbon dioxide levels might have melted the great ice sheets of the Pleistocene. He speculated, for the first time, on how carbon dioxide released into the atmosphere from the smokestacks of the Industrial Revolution would warm the planet. Like Fourier, he appears to have harbored a certain aversion to cold: he welcomed the greenhouse effect and looked forward to a warmer world.

For some time, no one worried much about climate change. The predictions were interesting but not relevant. It was a time of exploration and of scientific and technical revolution. The Franklin expedition, freezing and starving, probably resorted to cannibalism before perishing on a windswept gravel coast. Greely, after leaving a wake of corpses in the Arctic, failed to predict the School Children's Blizzard. Hydrogen was liquefied, and then helium, at only seven degrees Fahrenheit above absolute zero. Einstein and Bose described a new state of matter, a condensate that would exist only near absolute zero. Onnes discovered superconductivity at seven degrees Fahrenheit above absolute zero. Clarence Birdseye marketed frozen foods. There were other distractions: the rise of communism, a deadly flu epidemic, a world war.

And then, in the late 1930s, an English steam engineer named Guy Callendar found that the world's temperatures had been increasing over the past century. Digging a bit more, he found that carbon dioxide levels also had increased. But the historical data were suspect. Old thermometers had precision problems. Carbon dioxide measurements changed every time the wind changed direction. Conventional wisdom suggested that carbon dioxide would dissolve in the world's oceans. Callendar became a footnote in climatology textbooks.

By the 1950s, scientists were measuring atmospheric carbon dioxide more accurately. A gifted oceanographer named Roger Revelle was writing about climate change and sea level rise. His key point was that carbon dioxide unleashed from fossil fuels would not simply dissolve in the oceans and disappear, as had been previously believed. "Human beings," he wrote, "are now carrying out a large scale geophysical experiment of a kind that could not have happened in the past nor be reproduced in the future." Coupled with a new understanding of world population growth and its implications, the prophets warmed up to the idea that the world's temperature might increase as a result of human emissions. They did not like the world they envisioned. They began to worry.

For a time, climate change became something of a religious cause. One either believed or did not believe. The data were far from certain. The implications were huge. Industry representatives talked of climate change kooks and global warming crazies. Right-wing politicians ridiculed scientists. With each summer heat wave, the media recycled stories of a warming earth. With warming winters, the papers reran the stories from a slightly different angle. Often reporters botched the facts. People talked of a climate change debate, as if the reality of increased carbon dioxide and warmer temperatures could be discussed onstage, as if science could somehow be conducted under parliamentary rules.

The confusion left the science vulnerable. Naysayers pointed to events of fifty-five million years ago, to the Paleocene-Eocene Thermal Maximum, noting that the Arctic Ocean had warmed to seventy degrees, that global average temperatures had shot up five or ten degrees in a few thousand years, that the world had been hotter then than now. Yet the world had not ended. In fact, warming also had occurred in the midst of an ice age, in the heart of the Pleistocene. The climate change kooks and the global warming crazies pointed out that temperatures were rising quickly. Temperatures were rising much faster than they had fifty-five million years ago. The world, they said, did not end fifty-five million years ago, but habitats were flooded by rising sea levels. Animal and plant communities changed quickly. Today's warming would be more abrupt. And today we call those habitats cities. Among the plant communities likely to change quickly are fields of grain and corn and okra and broccoli. Among the animal communities likely to change quickly are those that include cows and chickens and humans.

The geologists, with a four-and-a-half-billion-year perspective, tended to camp with the naysayers. The biologists and the climatologists tended to camp with the climate change kooks. Over time, data and common sense made the kooks less kooky. For half a million years, carbon dioxide levels never passed three hundred parts per million, but the Industrial Revolution had sent them toward four hundred parts per million. Average temperature had risen a degree in a century. The risk that temperatures might keep rising was not acceptable. For the naysayers, the temperature rose and the warming tide turned. The heat-activated pendulum swung the other way and struck them squarely between the eyes. The kooks started to look wise. The naysayers started to look like kooks. Chemical companies started coming around. Oil companies started coming around. Car manufacturers started coming around. It suddenly seemed inevitable that the world would warm. Crops would fail as farms turned to desert. Sea level would rise as the remaining Pleistocene ice sheets melted away from Greenland and the snowfields and gla-ciers of Alaska and Canada turned to liquid. Long-term investments in low-lying Miami and Louisiana and the Maldives looked impru-dent. Polar bears drowned in the growing stretches of open water that scarred the Arctic sea ice. Their cubs appeared undernourished and slow growing. Four reports of bear cannibalism were attributed to warming temperatures and less ice. If nothing could be done, the white bear would go extinct.

Thanks to the Industrial Revolution, thanks to coal and oil and natural gas, thanks to more than six hundred million cars and trucks and buses, spring is fading into hothouse summer on planet Earth. Ten thousand years ago, spring arrived, and the Pleistocene Ice Age's most recent bout of extensive glaciation began to fade, but not so much as to melt the polar ice. Now the real demise of the Pleisto-cene Ice Age, full-blown summer, may be on the way.

The good news is this: the planet is not warming evenly. As ocean currents change, temperate Europe may become pleasantly frigid. And the Antarctic interior, surrounded by swirling winds thought to be driven in part by the hole in the ozone layer, has cooled. There will still be some use for Thinsulate and Gore-Tex, Hollofil and Quallofil. There will still be opportunities to wear a double layer of caribou skin.

And there is this: the naysayers have not given up. At the very least, they feel that the threat is overstated, and they challenge the connection between changing climate and human activities.

From Professor Bob Carter, Marine Geophysical Laboratory at James Cook University in Australia, with regard to Al Gore's popular movie about global warming: "Gore's circumstantial arguments are so weak that they are pathetic."

From Richard Lindzen, Alfred P. Sloan Professor of Meteorology at the Massachusetts Institute of Technology: "A general character-istic of Mr. Gore's approach is to assiduously ignore the fact that the earth and its climate are dynamic; they are always changing even without any external forcing. To treat all change as something to fear is bad enough; to do so in order to exploit that fear is much worse."

From Patrick Michaels, University of Virginia, on shifting animal and plant communities: "With all due respect, you would expect to see some slight changes in the distribution of plants and animals as the planet warms — or as the planet cools for that matter. It's hardly newsworthy."

❄ ❄ ❄

It is June ninth and fifty degrees in Anchorage. The caterpillars Fram and Bedford are dead. Their bodies, curled up in the bottom of their jar among dry leaves and branches, support a forest of fungus. The frozen mud I gathered, now thawed, is equally lifeless. The cold of the freezer may have been too sudden and perhaps too cold. Spring did not come for Fram and Bedford, nor for the frozen mud, any more than it is likely to come for James Bedford. Remove Bedford from his cryonic sepulcher filled with liquid nitrogen, lay him on a bed to warm, and all that is likely to result is a forest of fungus and the smell of death, a smell cryonically postponed but as inevitable as spring itself.

In a ten-page open letter addressed to James Bedford "and those who will care for you after I do," written in 1991 by a man describing himself as one of the caretakers of Bedford's frozen body, the author talks about moving Bedford's frozen remains from one unit to another. During the move, the author wrote, "we wrapped you in an additional sleeping bag, secured you in an aluminum 'pod' and transferred you to one of our new, state-of-the-art Dewars." It is not clear if the author of the letter understood that the Dewar in which James Bedford was suspended — little more than an oversize thermos bottle — was named after its inventor, James Dewar, who had been the first to liquefy hydrogen at 418 degrees below zero, almost a hundred degrees colder than the temperature at which James Bedford now resides. The author saw other advantages to the new Dewar thermos that held Bedford: "No more careening around the freeway every year or so" to have the unit serviced, and the new unit could store patients vertically, holding four of them together in a single Dewar thermos, taking up less space than Bedford's original horizontal unit had required.

From Dr. Ralph Merkle, a believer in cryonics as a means of avoiding the alternative, which, he points out, is certain death: "Cryonics is an experiment. So far the control group isn't doing very well."

From science and science fiction writer Arthur C. Clarke, during the early days of cryonics: "Although no one can quantify the probability of cryonics working, I estimate it is at least ninety percent—and certainly nobody can say it is zero."

From cryobiologist Dr. John Baust: "The individual who freezes himself or herself to come back in the future makes the assumption he will be a contributor to that society."

From cryobiologist Dr. Arthur Rowe, more adamantly a naysayer: "Believing cryonics could reanimate somebody who has been frozen is like believing you can turn hamburger back into a cow."

I telephone Alcor, a foundation that specializes in human and pet cryonics and that now looks after Bedford. Alcor refers to all its frozen wards as "patients." Alcor depends on members for financial support. In exchange, members have the option of postmortem preservation by freezing, with future resuscitation dependent on currently unknown means. That is, members have the option of becoming patients. A recorded message thanks me for calling. "If you are reporting the death or near death of an Alcor member," the recording says, "press two now."

I do not press two. Instead, I leave a message inquiring about Bedford's well-being, but I do not expect a return call.

❄ ❄ ❄

At the start of the twenty-first century, certain migratory birds seem to be showing up at the wrong places and the wrong times. Cranes that once wintered in Spain and Portugal now stop in Germany. For red knots, which nest in Siberia and winter in Africa, the expansion of African deserts may be the end of the road. In Australia, sandpipers, kingfishers, and plovers arrive two weeks earlier and stay three weeks later than they did in 1960. In the Shetland Islands, seven thousand pairs of skuas, starving because of a change in water temperature and the subsequent exodus of the fish they relied on for food, fail to rear chicks. According to the National Wildlife Federation, Maine will lose nineteen species to a warmer climate, including the olive-sided flycatcher, the boreal chickadee, and the dark-eyed junco, but it will gain eleven, including the Carolina chickadee, the Kentucky warbler, and the loggerhead shrike. The willow flycatcher and the black-capped chickadee may leave California, but they will be replaced by the cave swallow and the prothonotary warbler.

From Bert Lenten, executive secretary of the African-Eurasian Waterbird Agreement, on climate change: "Migratory birds are particularly vulnerable because of their use of several habitats during migration as stopover sites for feeding, resting, or to sit out bad weather."

From Robert Hepworth, executive secretary of the Convention on Migratory Species: "Species that adapted to changes over millennia are now being asked to make these adaptations extremely quickly because of the swift rise in temperatures."

From the World Wildlife Federation: "Birds are quintessential 'canaries in the coal mine' and are already responding to current levels of climate change."

Flowers and butterflies are canaries, too. Aspens are said to bloom twenty-six days earlier than they did a century ago. In the recent past, a Spaniard living in Barcelona could see the pretty little sooty copper butterfly in city gardens, but now he has to drive sixty miles north.

Glaciers sing like canaries. Fifteen hundred square miles of Alaska's Denali National Park are covered by glaciers. This is an area roughly equivalent to that of Rhode Island. The difference between the two: Rhode Island's area remains reasonably stable, while that of Denali's glaciers is shrinking. Compare photographs of Sunset Glacier in 1939 and 2004, and you will compare black-and-white mountainsides covered with snow and what must be blue ice to full-color mountainsides with patches of snow and no blue ice. Ditto for Mount Eielson photographs, Kahiltna Glacier photographs, and Muldrow Glacier photographs.

What the photographs do not show is that glaciers retreat in thickness as well as extent. Park scientists fly out to glaciers in carbon-emitting helicopters, use ice radar to measure the thickness, and shake their heads in disbelief. At Wrangell–St. Elias National Park, deep in Alaska's interior, where ice and snow cover an area larger than Connecticut, arrow shafts with intact feathers and spear points made from antlers pop up from newly thawed ground. At Glacier Bay National Park, along Alaska's southeast coast, the ground itself, relieved of the weight of snow and ice, rebounds at a rate of more than an inch each year.

The Beaufort Sea sings a song of less ice. In a second verse, it sings of openings in the pack ice that will be more abundant but less predictable than they were in the past. For narwhals, belugas, and bowhead whales hungry for a breath of air, this loss of predictability is not a pleasant song.

On Alaska's North Slope, when snow melts and then refreezes to form ice layers, as it sometimes does during warm winters, caribou that scratch out a living by hoofing aside snow may find themselves on unplanned diets. The same ice layers may trap voles and lemmings in their subnivean lairs.

Capitalism itself sings like a canary. From the chief executive of a clothing chain whose sales plummeted with rising temperatures: "All my analysis and all our data within the business is saying that it's a weather thing."

From Bill Ford Jr., executive chairman of Ford Motor Company: "We see climate change as a business issue as well as an environmental issue and we're accelerating our efforts to find solutions."

From an international oil company: "Greenhouse gas levels are rising and the balance of scientific opinion links that rise to the increase in our planet's surface temperatures. As a major provider of energy, we believe we have a responsibility to take a lead in finding and implementing solutions to climate change."

And from Alexander Cockburn in the June 9, 2007, edition of *CounterPunch*: "Capitalism is ingesting global warming as happily as a python swallowing a piglet."

❄   ❄   ❄

It is June nineteenth and fifty-five degrees on Alaska's North Slope. I drive across the oil fields. The larger lakes remain half-covered with ice, but the only remaining snow stands in piles deposited by plows. The piles are filthy, laced with dust and road gravel.

The landscape, though snow-free, remains brown. Arctic foxes, their coats mangy as they turn from winter white to summer brown, run across tundra soaked by meltwater. Skinny caribou, back from the migration and shedding winter fur, appear mangy, too. I pass king eiders and loons in now liquid ponds and hundreds of white-fronted geese staggering around on the tundra. Groups of ten or twenty snow geese forage along the edge of a road near the coast. Over the next twelve weeks, their chicks will hatch. The flightless hatchlings and their parents will march fifteen miles east across the tundra to feed in salt marshes. As summer progresses, the goslings will grow and fatten up and learn to fly. They will head south just as the snow starts to accumulate, staying ahead of the short days and the bitter winds.

On the edge of a gravel road, I listen to a biologist describe his experiments with hibernating ground squirrels. "They look dead," he says. "They are curled up. If you uncurl them, and they are alive, they will curl up again. If they are dead, they will not curl up again." He collects hibernating animals from the field and takes them back to his laboratory. While they hibernate, he inserts probes in their brains. He measures their temperature. Funding, he says, comes from the military and the medical community. In addition to being world-class hibernators, they can, for reasons not fully understood, survive massive blood loss. Behind him, out of his line of sight on the tundra, a ground squirrel stands on its hind feet, uncurled and vigilant, posing like a prairie dog. Until a few weeks ago, this squirrel had been nearly frozen, curled up. Farther behind, the Putuligayuk River flows gently, ice-free, the sudden flush of breakup already over. Between me and the river, wooden piles, left over from a long-abandoned drilling operation, have been frost-heaved out of the ground. Little mounds of gravel cover the tops of the piles several feet above the ground. Between the piles with their mounded gravel, male plovers dance about, spreading their wings and hopping above sprigs of grass, hoping to attract females. It has been said that they fly halfway around the world with no greater purpose in mind than sex. It might be better said that they fly halfway around the world with no lesser purpose in mind than sex. With sex out of the way, they stay with the females and the young, sharing the responsibilities of nesting, egg sitting, feeding of young, and flight school.

I drive toward the coast to meet a group intent on releasing a seal. The seal was found inland months ago with its jaw broken and its flippers and face frostbitten. It was captured, flown south, and treated. Today it is to be set free. It shows up in a plastic dog kennel of the sort used to fly pets. Two biologists carry the kennel from the back of a truck and place it on a gravel beach. An oil field worker wearing a hard hat and an industry-issued fireproof work jacket unlatches the door. The seal, seemingly suspicious, sniffs the air. It shuffles a flipper length forward and looks around. It shuffles another few flipper lengths and stops to look around again. Its eyes are brown spheres of the sort that would melt the heart of the most field-hardened biologist. Its body is that of a Butterball turkey, almost as wide as it is round. An orange satellite transmitter has been glued to its back and will stay there until it sheds its fur next spring. It breaks for the sea, shuffling hastily across the remaining few feet of gravel. It plunges into a light chop. It swims, then pauses. Head up, it looks out toward the sea ice, perhaps a mile to the north. It turns to bless us once more with its brown-eyed gaze, then submerges, dipping like a submarine. When it reappears a few minutes later, it is heading due north, toward the sea ice. Toward home.

❄ ❄ ❄

Warm water is less dense than cold water, and hence warm water floats on top of cold water. Freshwater is less dense than salt water, and hence freshwater floats on top of salt water. Water warmed in the tropics floats on the colder underlying water and spreads out to the north. As it spreads, it loses heat. A portion of it evaporates, and it becomes increasingly salty. By the time it reaches the North Atlantic, it has lost enough heat and become salty enough to sink. The sinking water spreads south across deep ocean basins. It may not rise again for a thousand years.

This movement of ocean water has been called the conveyor belt. The belt conveys the equivalent of one hundred Amazon Rivers and carries enough heat to warm western Europe. Without this belt, Berlin might be as cold as Edmonton. Without this belt, Scottish sheep might be grazing hip-deep in snow.

Great Britain's former prime minister Tony Blair, advised by scientists, began to worry about sudden changes in climate, about what more and more specialists were thinking of as catastrophic tipping points. One thing he might have had on his mind was a conveyor belt breakdown. As polar ice melts, it dumps freshwater into the North Atlantic. North Atlantic water is cold and salty, so it sinks, and in sinking it powers the conveyor belt. Melted ice is not salty. It will not sink as well as the cold salty water of the North Atlantic. By not sinking, it will slow the conveyor belt. With the greenhouse effect and global warming, western Europe will warm for a time, but if the conveyor belt slows, western Europe may suddenly cool down. British farmers, it has been said, may have to learn to grow crops in snow. Blair may have been worried, too, about the tipping point of the ice itself. For the three decades since satellites have been in place to watch the poles, more than a quarter of the once permanent pack ice has disappeared. The melting of ice, the weakening of the hydrogen bonds that hold water molecules in the lockstep crystalline structure of ice, requires a great deal of heat, but that heat does not change the temperature of the ice or water. The heat does nothing more than weaken the hydrogen bonds between molecules. A one-square-mile block of ice measures thirty-two degrees just before it melts. Apply heat until the one-square-mile block of ice melts, and the pool of liquid water will measure thirty-two degrees. Apply the same amount of heat needed to melt that block of ice to the pool of liquid water and the water temperature will rise one hundred and seventy-six degrees. After the ice is gone, after the hydrogen bonds are weakened, after the solid collapses into a formless liquid: add heat and the temperature soars.

And Blair may have worried, too, about methane. As permafrost melts, as Arctic tundra warms, the methane trapped in soil will be released into the atmosphere. Methane is twenty times more effective at trapping heat than carbon dioxide. In 2005, scientists looking at Russian permafrost reported melting. What had been frozen peat was becoming a landscape of mud and lakes. An area the size of Germany and France combined could be poised to release seventy billion tons of methane. More would come from Alaska and Canada and various mountain peaks. Sergei Kirpotin at Tomsk State

University in western Siberia called the situation "an ecological land-slide." David Viner, a senior scientist at the University of East Anglia in England and part of the Russian permafrost project, said, "There are no brakes you can apply."

❄   ❄   ❄

It is June twentieth, the eve of the summer solstice, on Alaska's North Slope and 60 degrees Fahrenheit, or 16 degrees Celsius, or 289 Kelvin. Said another way, it is 520 degrees Fahrenheit above absolute zero.

A man tells me of a musk ox frozen in the sea ice, standing up, thirty miles west of here. I find this hard to believe. He produces a photograph. The ox is indeed standing in ice and frozen. Its head hangs low, in the normal posture of a musk ox, but its breath has formed a frozen pedestal. It reminds me of the stories of cattle frozen during the School Children's Blizzard. The difference is that here the ox will fall into the Beaufort Sea as the ice melts. The man with the photograph is worried that the carcass, melting, will attract polar bears.

I stand on the western edge of Prudhoe Bay at the spot where I swam almost a year ago, on the first of July. Here the ice remains hard up against the shore. It is rotten ice, very nearly the kind of ice called *aunniq* by the Inupiat, but without holes of open water. Swimming is out of the question anywhere nearby. The ice stretches out across Prudhoe Bay and to the north, ragged and in places dirty, in places cracked, in places heaved up into miniature pressure ridges. Brownish blocks of ice stand above the frozen surface like glacial erratics. The blocks cast shadows that look like seals, but I see no actual seals.

I think for a moment of Father Henry living in his ice cave, of the caterpillars just becoming active on the tundra, of Adolphus Greely. Tomorrow will mark the anniversary of Greely's salvation. Sergeant

David Brainard, one of the survivors of Greely's expedition, did not record a journal entry on June twentieth. He did, however, write an entry for the twenty-first, about nine hours before the rescue ship sailed into view: "Our summer solstice! The wind is still blowing a gale from the south. Temperature 7 a.m. 31°, minimum recorded 28°." He ate lichen stew and boiled sealskin that day. "Since the day before yesterday," he wrote, in a comment that would be echoed in the journal of his savior, "Elison has eaten his stew by having a spoon tied to the stump of his frozen arm." At eight thirty in the evening, the rescue ship steamed into sight. Brainard, Greely, and four others returned to civilization and family and careers. Elison, taken aboard the rescue ship but too far gone to be saved, died in Greenland on July 8, 1884, just seventeen days after the rescue.

At this moment, there is just enough breeze to keep the mosquitoes down. I consider applying sunscreen to my ears and nose. I look out across the sea ice. Through binoculars, I have to look well offshore to find pockets of open water. There will be no walking to the North Pole today, but the arrival of a rescue ship would also be out of the question.

With a perverse desire to swim, I consider driving to the other side of Prudhoe Bay, to the open water where the seal was released. But I decide against it. Instead, I stand next to the ice, worn out from a long winter, enjoying the breeze and the view of the sea ice, simultaneously disappointed and thankful that it remains too cold for swimming.

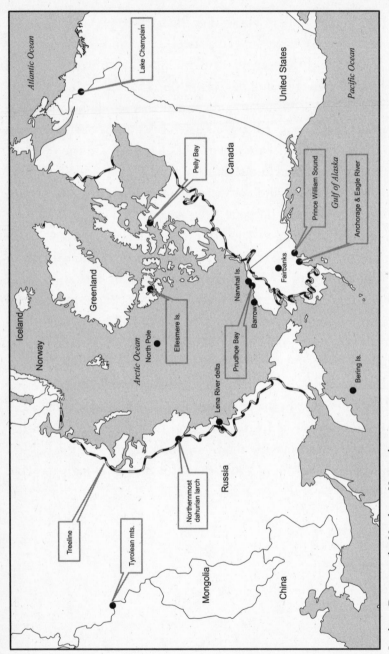

Looking Down on the Northern Hemisphere
Geographic illustration prepared by Bill Lee

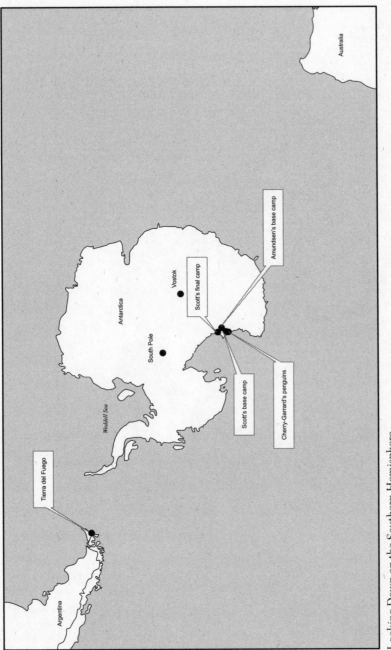

Looking Down on the Southern Hemisphere

Geographic illustration prepared by Bill Lee

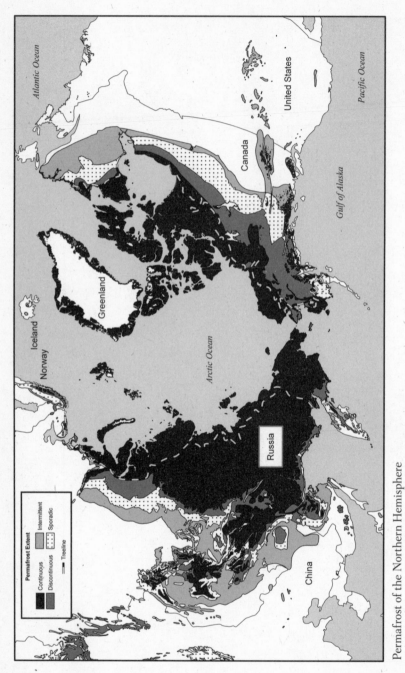

Permafrost of the Northern Hemisphere

Geographic illustration prepared by Bill Lee based on J. Brown, O. J. Ferrians Jr., J. A. Heginbottom, and E. S. Melnikov. 1997. Circum-Arctic map of permafrost and ground-ice conditions. U.S. Geological Survey Circum-Pacific Map CP-45, 1:10,000,000. Reston, Virginia.

ACKNOWLEDGMENTS

Over the past eight years, I have met and worked with hundreds of people in the Alaskan Arctic. Inupiat hunters, biologists, archaeologists, oil field workers, activists, engineers, teachers, and others have all influenced my thinking about the Arctic and about cold in general, as have the many gifted authors who have written about the world's cold regions. My agent, Elizabeth Wales, worked hard to place this book with Little, Brown and Company. My editor, John Parsley, provided important comments and suggestions that improved the book, while also patiently walking me through the publication process. After I thought the book was finished, copyeditor Barb Jatkola pointed out hundreds (literally) of opportunities for improvement. Glenn Wolff, with talent and tolerance of my vagaries, drew illustrations that captured key aspects of the text. Although I have never met Robert Twigger, his book *The Extinction Club* inspired the approach I used in *Cold*. Lisanne Aerts, Matt Cronin, Jason Hale, John Kelley, Amy King, Bill Lee, and Kathryn Temple provided comments on the draft manuscript as well as much-needed encouragement. Bill Lee, John Kelley, and especially Lisanne Aerts also appeared, along with many others, as unnamed companions in various passages in the book. Lastly, my dog, Lucky, deserves credit for his willingness to lie on the office floor while I worked, even though he would rather have been out in the snow.

NOTES

With a Few References, Definitions, Clarifications, and Suggested Readings

JULY

The words "Inupiat," "Iñupiaq," "Inupiaq," and "Inupiak" are sometimes used interchangeably, although some residents of the far north say that "Inupiat" refers to the people while "Iñupiaq" refers to the language, or that "Inupiat" should be used as a noun and "Iñupiaq" as an adjective (as in "Iñupiaq people"). The Inupiat include some of the Alaskan native coastal people, or Alaskan Eskimos, one of many Inuit, or native coastal people of the Arctic in Alaska, Canada, and Greenland. Iñupiaq (the language) is notoriously difficult for outsiders, but online dictionaries are available, such as the *Iñupiaq Eskimo Dictionary* at www.alaskool.org/language/dictionaries/inupiaq.

World War II, like many other wars, provides interesting stories of hypothermia and frostbite. The story of the foundered German troop carrier is related in R. Tidow's "Aerzliche Fragen bei Seenot" (1960, *Wehrmedizinische Mitteilungen*), which was summarized in an attachment to Lorentz Wittmers and Margaret Savage's "Cold Water Immersion," published as chapter 17 in volume 1 of *Medical Aspects of Harsh Environments* (2002, Department of the Army, Office of the Surgeon General, Borden Institute). *Medical Aspects of Harsh Environments* also includes an atlas of cold-related injuries with gruesome but interesting full-color photographs of frostbitten hands, feet, ears, and noses.

The book can be viewed at www.bordeninstitute.army.mil/published_volumes/harshEnvi/harshEnvi.html.

Orcutt Frost's *Bering: The Russian Discovery of America* (2003, Vail-Ballou Press, New York) was the first book-length biography of Bering in a hundred years. Neither Bering's life itself nor written descriptions of his life have been kind, but his journeys across Siberia and then to America were remarkable accomplishments.

Adolphus W. Greely is a well-known figure. His memoir, *Three Years of Arctic Service: An Account of the Lady Franklin Bay Expedition of 1881–84 and the Attainment of the Farthest North,* was first published in 1886 (Charles Scribner's Sons, New York). Reprints remain available on the used-book market.

Father Henry is described in *Kabloona: Among the Inuit*, written by the French aristocrat Gontran De Poncins and Lewis Galantiere in 1941 and recently reprinted as part of the Graywolf Rediscovery Series (1996, Graywolf Press, St. Paul). The word "Kabloona" means "white man." The book tells the story of Father Henry, and also the story of one of the author's understanding and to some degree accepting the ways of a very foreign culture.

Tom Shachtman's *Absolute Zero and the Conquest of Cold* (1999, Mariner Books, New York) gives an engaging account of the history of thermometers. Daniel Fahrenheit, whose surname has been immortalized by his temperature scale, contributed one step in the long and still ongoing journey of temperature measurement. His scale is not especially useful outside the ranges typically encountered by humans.

Although Celsius can be converted to a rough approximation of Fahrenheit by multiplying by two and then adding thirty-two, a more exact conversion is degrees Fahrenheit = (degrees Celsius × 9/5) + 32. Similarly, Fahrenheit can be converted to a rough approximation of Celsius by subtracting thirty-two and then dividing by two, but a more exact conversion is degrees Celsius = (degrees Fahrenheit − 32) × 5/9.

NOTES

The precise conversion of zero Kelvin, or absolute zero, is usually given as 459.67° below zero Fahrenheit.

The complete quotation about achieving a temperature low enough to result in the formation of a super atom, from Eric Cornell, as reported in a joint press release by the University of Colorado and the National Institute of Standards and Technology on July 13, 1995, was "This state could never have existed naturally anywhere in the universe. So the sample in our lab is the only chunk of this stuff in the universe, unless it is in a lab in some other solar system." The experiment, credited jointly to Cornell and Carl Wieman, had taken six years and involved eight graduate students and three undergraduate students. While talking to a reporter about the Nobel Prize that came from this work, Wieman explained that he had to rush off to teach a physics class for nonscientists. Despite his success, he retained the dedication and modesty needed to teach undergraduate physics to a broad range of students.

Apsley Cherry-Garrard, of the Scott expedition, seemed to especially enjoy talking about temperature in terms of "degrees of frost." He often talked of "degrees of frost" in is his 607-page memoir, *The Worst Journey in the World* (reprint, 2000, Carroll and Graf, New York), which has won high praise as an example of fine travel writing. Polar explorers were not the only ones to measure cold temperatures in degrees of frost. In the famous short story "To Build a Fire," Jack London refers to a temperature of "one hundred and seven degrees of frost."

Dante's *Inferno* can be described as a travelogue through the circles of Hell. Canto Thirty-one takes readers into the tenth and final circle of Hell, where traitors reside, including "betrayers of kindred" who have murdered their brothers, as well as Judas Iscariot, Brutus, and Cassius. In Canto Thirty-four, still in the frozen tenth circle of Hell, readers meet Satan himself: "The emperor of the reign of misery from his chest up emerges from the ice."

Gynaephora rossii, the woolly bear caterpillar and the moth that it becomes, is described by D. C. Ferguson in a chapter called "Noctuoidea (in part): Lyantriidae" in the 1978 book *The Moths of America North of Mexico,* edited by R. B. Dominick et al. (E. W. Classey, London).

Neil Davis's textbook *Permafrost: A Guide to Frozen Ground in Transition* (2001, University of Alaska Press, Fairbanks) offers a thorough technical introduction to permafrost and the formation of various permafrost features, such as pingos, ice wedges, polygonized ground, and frost boils. Davis has been criticized for undertaking a textbook outside his own field of geophysics, but nevertheless *Permafrost* is well worth reading.

The story of the abandonment of the *Endurance* is well known, mostly because of Ernest Shackleton's *South: The* Endurance *Expedition,* which has been widely read since its original publication in 1919 (William Heinemann, London). The book was reprinted in 1999 by Signet (New York).

Charles Wright's interviewer was Charles Neider, who edited the book *Antarctica: Firsthand Accounts of Exploration and Endurance* (2000, Cooper Square Press, New York). His interview of Wright appears in a chapter called "Beyond Cape Horn" in the collection *Ice: Stories of Survival from Polar Exploration,* edited by Clint Willis (1999, Thunder Mouth Press, New York).

Robert Falcon Scott's journals were reprinted in 1996 as *Scott's Last Expedition: The Journals* (Carroll and Graf, New York).

Captain George E. Tyson's *Tyson's Wonderful Drift* was published in 1871 and is now difficult to find. However, it has been reprinted in part in the collection *Ring of Ice: True Tales of Adventure, Exploration, and Arctic Life* (2000, Lyons Press, New York).

Roald Amundsen is sometimes described as the most practical of the polar explorers. He considered "adventure" to be "an unwelcome interruption" of the explorer's "serious labours," and he was critical of the poor planning and poor judgment that so often led to tragedy during exploration. His book *Roald Amundsen—My Life as an Explorer* (1927,

23

Doubleday, Garden City, NY) came out one year before he disappeared when his plane crashed during a rescue mission in the Arctic.

Various versions of *The Story of Comock the Eskimo* remain available, including one published by Simon and Schuster (New York) in 1968. The authors are Comock, R. J. Flaherty, and E. S. Carpenter.

Frederick Albert Cook's *My Attainment of the Pole* (1913, Mitchell Kennedy, New York) was reprinted in 2001 by Polar Publishing (New York). Cook, considered a charlatan by many of his contemporaries, believed that few people "have ever been made to suffer so bitterly and so inexpressively as I because of the assertion of my achievement."

De Long died in Siberia, but his widow, Emma J. Wotten De Long, edited and published his journal entries under the title *The Voyage of the* Jeannette: *The Ship and Ice Journals of George W. De Long, Lieutenant-Commander U.S.N., and Commander of the Polar Expedition of 1879–1881* (1884, Houghton Mifflin, New York).

The origin of the name of Narwhal Island does not seem to be well documented, but during a lunchtime conversation in 2008 in Anchorage, an Inupiat hunter told me that his grandfather had sailed on the *Narwhal*. The hunter thought the ship might have used the island as a base during the whale hunt, probably in the late 1800s.

Adolphus W. Greely's quotations about the wretched conditions of camp life come from his memoirs, published by Charles Scribner's Sons (New York) in 1886 as *Three Years of Arctic Service: An Account of the Lady Franklin Bay Expedition of 1881–84 and the Attainment of the Farthest North*.

David L. Brainard's account of the disastrous Greely expedition was published in 1940 as *Six Came Back: The Arctic Adventure of David L. Brainard* (Bobbs-Merrill, Indianapolis). Brainard, the last survivor of the Greely expedition, died in 1946.

The scientific paper that estimated the caloric needs of the Greely expedition was "Chances for Arctic Survival: Greely's Expedition

Revisited," written by Jan Weslawski and Joanna Legezynska and pub-lished in the journal *Arctic* (2002, vol. 4, pp. 373–79).

W. S. Schley and J. R. Soley's *The Rescue of Greely* (1885, Charles Scrib-ner's Sons, New York) describes the rescue of Greely but also gives a vivid description of routine life aboard vessels sailing to the Arctic and the challenges they faced even when things went well.

Windchill is a well-known concept today, but the original report by Paul Siple and Charles Passel on their work quantifying the effect of windchill was not available until 1945, when it was published in the *Proceedings of the American Philosophical Society* (vol. 89, pp. 177–99) as "Measurements of Dry Atmospheric Cooling in Sub-freezing Tem-peratures; Reports on Scientific Results of the United States Antarctic Service Expedition, 1939–1941."

For a remarkably detailed and engaging account of the Blizzard of 1888, read David Laskin's *The Children's Blizzard* (2004, Harper Perennial, New York).

AUGUST

Admiral Richard E. Byrd's 1938 memoir, *Alone: The Classic Polar Adventure* (reprint, 2003, Island Press, Washington, DC), was a best seller when first published. For today's readers, it remains the story of a man working alone in isolation—a story of self-discipline and tough-ness of spirit, body, and mind.

An Arctic Boat Journey: In the Autumn of 1854 (1871, James A. Osgood and Co., Boston), by Isaac I. Hayes, was republished in 2007 by Kes-singer Publishing (Whitefish, MT). Kessinger Publishing digitizes rare books, including many about the Arctic. Also available at www.archive.org/details/arcticboat00hayerich.

Fridtjof Nansen's *Farthest North: Being the Record of a Voyage of Explo-ration of the Ship* Fram *1893–1896, and of a Fifteen Months' Sleigh Journey by Dr. Nansen and Lt. Johansen* was originally published in Norwegian in 1897. It has since been republished many times, including a 2008 version published by Skyhorse Publishing (New York) under the title *Farthest North: The Epic Adventure of a Visionary Explorer*. The *Fram* is a ship with an amazing history. It has been preserved at the Frammuseet, or Museum of the *Fram*, on Bygdøy Island in Oslo, Norway.

For photographs and a history of the permafrost tunnel, see www.crrel .usace.army.mil/permafrosttunnel/. Although the tunnel is not open to the public, tours are sometimes arranged for visiting students, engineers, and scientists. An admirable technical paper titled "Syngenetic Permafrost Growth: Cryostratigraphic Observations from the CRREL Tunnel Near Fairbanks, Alaska," by Y. Shur, H. M. French, T. Bray, and D. A. Anderson (2004, *Permafrost and Periglacial Processes*, vol. 15, pp. 339–47), gives detailed descriptions of the permafrost tunnel's features.

Permafrost is usually defined as ground that remains at temperatures below thirty-two degrees for two years or longer. Ice may or may not be present. For example, dry bedrock in the far north may not contain ice. The permafrost zone is often divided into "continuous permafrost" and "discontinuous permafrost," and sometimes further divided into "intermittent" and "sporadic."

The steppe bison recovered near Fairbanks was dubbed Blue Babe after Paul Bunyan's ox because specks of blue iron phosphate (vivianite) dotted its skin. It was preserved through the efforts of Dale Guthrie and remains on display at the University of Alaska's museum in Fairbanks. For a more detailed description, see Mary Lee Guthrie's well-illustrated *Blue Babe: The Story of a Steppe Bison Mummy from Ice Age Alaska* (1988, White Mammoth, Fairbanks).

We are accustomed to the properties of water through everyday experiences, but water is an amazing compound. Water is the only nonmetallic substance known to expand when it freezes. Volume increases by about nine percent when water freezes and then decreases slightly as the temperature drops further. Marianna Gosnell celebrates the properties of water in her book *Ice: The Nature, the History, and the Uses of an Astonishing Substance* (2005, Alfred P. Knopf, New York).

Taiki Katayama and a number of coauthors (Michiko Tanaka, Jun Moriizumi, Toshio Nakamura, Anatoli Brouchkov, Thomas Douglas, Masami Fukuda, Fusao Tomita, and Kozo Asano) described growth of bacteria from an ice wedge in the permafrost tunnel in an article called "Phylogenetic Analysis of Bacteria Preserved in a Permafrost Ice Wedge for 25,000 Years" (2007, *Applied and Environmental Microbiology*, vol. 73, pp. 2360–63). They wrote, "Our results suggest that the bacteria in the ice wedge adapted to the frozen conditions have survived for 25,000 years."

No one should visit Fairbanks without reading Terrence Cole's engaging history of early mining, *Crooked Past: The History of a Frontier Mining Camp; Fairbanks, Alaska* (1991, University of Alaska Press, Fairbanks).

Between 1979 and 2002, 16,555 hypothermia fatalities were reported in the United States, with Alaska, New Mexico, North Dakota, and Montana having the highest number of hypothermia deaths in 2002. Most of these victims died before they could be treated. The number of victims who die from rewarming shock is not known.

Mechem's article called "Frostbite," available at www.emedicine.com/emerg/topic209.htm, is intended as a quick reference for medical professionals, but it includes information sure to interest anyone who travels in cold regions. It also includes interesting photographs of badly frostbitten hands, ears, and feet.

The article "Modified Cave Entrances: Thermal Effect on Body Mass and Resulting Decline of Endangered Indiana Bats (*Myotis sodalist*)," by A. R. Richter, S. R. Humphrey, J. B. Cope, and V. Brack, appeared in the academic journal *Conservation Biology* (1993, vol. 7, pp. 407–15). Anyone who has spent time in caves will intuitively understand that modifying entrances will change airflow and, subsequently, underground climates.

As Admiral Richard E. Byrd's 1938 book *Alone: The Classic Polar Adventure* (reprint, 2003, Island Press, Washington, DC) progresses, the battle with carbon monoxide poisoning and its effect on his mind becomes increasingly important.

Frederick Albert Cook's description of his meal on the pole, from his 1913 book *My Attainment of the Pole* (Mitchell Kennedy, New York), will ring true to anyone who has spent enough time in the backcountry to begin losing significant amounts of body fat. On long backcountry trips, one often spends the first few days feeling pleasantly tired but easily satiated, but as body fat disappears, one gradually begins to feel almost constant hunger, even after a large meal.

Part of Captain George E. Tyson's 1871 book *Tyson's Wonderful Drift* was reprinted in the collection *Ring of Ice: True Tales of Adventure, Exploration, and Arctic Life* (2000, Lyons Press, New York).

In their 1941 book *Kabloona: Among the Inuit,* reprinted as part of the Graywolf Rediscovery Series (1996, Graywolf Press, St. Paul), Gontran De Poncins and Lewis Galantiere suggest that consumption of frozen fish will keep one warm in cold climates. More commonly among the Alaskan Inupiat, walrus meat is said to have this property.

Bernd Heinrich's *Mind of the Raven* and *Ravens in Winter* are perhaps better known than his equally delightful book on winter ecology, *Winter World* (2003, HarperCollins, New York). Heinrich's quotations here are from *Winter World,* a book that should be read and reread by anyone living in a cold climate.

SEPTEMBER

The quotations and some of the background information on the Little Ice Age are from Brian Fagan's interesting book *The Little Ice Age: How Climate Change Made History, 1300–1850* (2000, Basic Books, New York). This book, more than any other, brings home the reality of even relatively minor changes in climate and reminds us that climate variability and predictability are as important as or more important than temperature itself. Additional information on the Little Ice Age can be found in dozens of books and articles.

The English vicar's quotation is from Henry Stommel and Elizabeth Stommel's *Volcano Weather: The Story of 1816, the Year Without a*

Summer (1983, Seven Seas Press, Newport, RI), which explains the difficulties surrounding the association of cold weather with volcanic eruptions.

Confusion surrounding the Pleistocene Ice Age is surprising. Also called the Quaternary glaciation and the Pleistocene glaciation, it is sometimes defined as the period during which permanent ice sheets existed in Antarctica and possibly Greenland, with fluctuating ice sheets and glaciers in other areas. During this period, temperatures fluctuated enough to allow large-scale expansion and contraction of ice sheets, or glacial and interglacial periods. It seems that many people think of the Pleistocene Ice Age incorrectly as equivalent to the Wisconsin glaciation, which lasted from about one hundred thousand years ago until about ten thousand years ago. The Wisconsin glaciation was one name for one of at least four periods of Pleistocene glaciation. The Pleistocene nominally ends, quite artificially, near the beginning of recorded human history, around ten thousand years ago, at a time that corresponds with the beginning of the current interglacial period.

The Mayo Clinic provides an excellent description of Raynaud's disease at http://www.mayoclinic.com/health/raynauds-disease/DS00433. The disease afflicts five to ten percent of people. Women are five times more likely than men to suffer from Raynaud's disease.

Darwin's comment lamenting his failure to spot obvious signs of past glaciation comes from *The Autobiography of Charles Darwin,* which has been republished many times, including in 1993 (W. W. Norton, New York). Even today, people trained in periglacial geology are far more adept at spotting signs of previous glaciation than those with only a passing knowledge of the topic.

Doug Macdougall's *Frozen Earth: The Once and Future Story of Ice Ages* (2004, University of California Press, Berkeley) gives an insightful and readable history of ice ages, including the history of the science of ice ages, with well-deserved emphasis on Agassiz.

An electronic copy of James Croll's 1875 book *Climate and Time in*

Their Geological Relations: A Theory of Secular Changes of the Earth's Climate (Adam and Charles Black, Edinburgh) is available at http://books.google.com/books?id=Q98PAAAAIAAJ&printsec=frontcover&dq=croll#PPP2,M1.

As recently as forty years ago, before the reality of global warming was widely recognized, serious scientists pondered the problem of the end of the current interglacial period and renewed cooling. Plans for preventing the return to a glacial period were discussed in earnest, including the possibility of intentionally scattering coal dust on the ice caps to melt the ice and end the reflection of heat back into space. An April 28, 1975, article in *Newsweek* described "a drop of half a degree [Fahrenheit] in average ground temperatures in the Northern Hemisphere between 1945 and 1968" and reported on "ominous signs that the Earth's weather patterns have begun to change." These discussions and the media attention they drew undoubtedly delayed the acceptance of data suggesting that greenhouse gas emissions are contributing to warmer temperatures.

It is often possible in Alaska to see the evidence of recent Pleistocene glaciation confounded by the evidence of older Pleistocene glaciation. It can be amusing to listen to scientists debate glaciation patterns in the field based on evidence they can observe from the edge of the road. They often use words such as "obvious" and "self-apparent" as they contradict one another's ideas. The further back one goes in time, the more speculative and obscure the evidence becomes.

Gabrielle Walker's wonderful *Snowball Earth: The History of a Maverick Scientist and His Theory of the Global Catastrophe That Spawned Life as We Know It* (2003, Three Rivers Press, New York) explains the scientific and human history of the Snowball Earth theory and provides a nice sketch of Paul Hoffman.

Paul K. Feyerabend's *Against Method* (1993, Verso, London) argues that the strength of personalities may be as important as data to the successful advancement of scientific ideas.

NOTES

OCTOBER

The description of the job opportunity was reported in "NASA Offers $5000 a Month for You to Lie in Bed," by Alexis Madrigal (May 7, 2008, *Wired Science,* http://blog.wired.com/wiredscience/2008/05/nasa-offers-500.html).

Much of the discussion of hibernation and other aspects of winter ecology stems from work described in Bernd Heinrich's *Winter World* (2003, HarperCollins, New York). Additional excellent information on hibernation and other winter adaptations comes from Peter Marchand's *Life in the Cold* (1996, University Press of New England, Hanover, NH) and James Halfpenny and Roy Ozanne's *Winter: An Ecological Handbook* (1989, Johnson Books, Boulder, CO).

Today technical publications on ecological studies are almost without exception steeped in complex statistical analyses, but Edmund Jaeger's "Further Observations on the Hibernation of the Poor-Will" (1949, *Condor,* vol. 51, pp. 105–9) provides an example of the sort of natural history that was the foundation of the science of ecology—hard-won observations from the field backed up by orderly thinking.

The two University of Minnesota researchers were J. R. Tester and W. J. Breckenridge. They described their work in a 1964 article titled "Winter Behavior Patterns of the Manitoba Toad, *Bufo hemiphrys,* in Northwestern Minnesota" (*Annales Academiae Scientiarum Fennicae,* series A. IV, *Biologica,* vol. 71, pp. 424–31). One cannot help but wonder what their families and nonbiologist friends thought when the two researchers explained that they were tracking toads, but biologists often find themselves explaining their work to incredulous nonspecialists.

John Burroughs's description of his discovery of a hibernaculum comes from his 1886 book chapter "A Sharp Lookout," which is available online at http://kellscraft.com/Burroughs,John/SignsandSeasons/SignsandSeasonsCh01.html. The full book, *Signs and Seasons,* was published by Riverside Press (Cambridge, MA).

William Schmid's work was published in 1982 as "Survival of Frogs in

Low Temperature" in the prestigious journal *Science* (vol. 215, pp. 697–98). Among other things, Schmid's article says that higher levels of glycerol during winter are related to frost tolerance. One wonders what the nineteenth-century naturalist John Burroughs would have thought about in-depth physiological work on frozen frogs.

The passage from Lynn Rogers's "A Bear in Its Lair" (October 1981, *Natural History*, pp. 64–70) describing his bear den encounter has been reprinted in full and in part many times. Although Rogers eloquently describes his experience, many bear biologists have similar stories. They work casually with large and potentially dangerous animals and are occasionally reminded of their own mortality.

The Pennsylvania State University student posted a description of his experience at the Barrens at http://lamar.colostate.edu/~benedict/weather/barrens.shtml.

Within Pennsylvania, there are other barrens: the Moosic Mountain Barrens and the Serpentine Barrens are two examples. John McPhee's 1967 article "The Pine Barrens," published in the *New Yorker* and later republished as a short book (1968, Farrar, Straus and Giroux, New York), describes life in one of America's more accessible barrens, in New Jersey. Many of the barrens scattered across the United States have been overcome by suburban sprawl, an unfortunate situation, since barrens often support unique plant and animal communities.

Scott Weidensaul's contribution to the avian literature *Living on the Wind: Across the Hemisphere with Migratory Birds* (1999, North Point Press, New York) provides useful information on bird migration. The quotations from Magnus, Aristotle, and Homer are cited in Jean Dorst's *The Migrations of Birds* (1962, Houghton Mifflin, Boston) and Weidensaul's *Living on the Wind*.

Bird collisions are an important conservation issue. Collisions with windows may kill more than one hundred million birds each year. An additional fifty million to one hundred million are killed in collisions with cars and trucks. Bird collisions with aircraft also pose a serious problem for both birds and aircraft. For example, a 1995 crash of an AWACs battlefield radar plane was attributed to collisions with geese, and a 1998 commercial jet flight made an emergency landing after experiencing an engine malfunction caused by a bird strike. In 2009, as *Cold* was going to press, another commercial flight was forced to make an emergency landing—this time in the Hudson River—when a bird strike caused engine failure. USA Bird Strike Committee maintains records, including photographs, of aircraft-bird collisions.

The estimate of one hundred million birds per year killed by cats in the United States is provided by the National Audubon Society. Daniel Klem of Muhlenberg College has estimated that cats kill seven million birds each year in Wisconsin alone.

NOVEMBER

The 2004 article on damselfish enzymes, by Glenn C. Johns and George N. Somero, is called "Evolutionary Convergence in Adaptation of Proteins to Temperature: A_4-Lactate Dehydrogenases of Pacific Damselfishes (*Chromis* spp.)" (*Molecular Biology and Evolution*, vol. 21, no. 2, pp. 314–20). Although the article is highly specialized, it is to some degree accessible to anyone with basic biological training.

The discussion about the cod fishery and other events of the Little Ice Age draws on Brian Fagan's *The Little Ice Age: How Climate Change Made History, 1300–1850* (2000, Basic Books, New York).

Gabrielle Walker, in her book *Snowball Earth: The History of a Maverick Scientist and His Theory of the Global Catastrophe That Spawned Life as We Know It* (2003, Three Rivers Press, New York), describes Douglas Mawson's book as "one of the best books ever written about Antarctic exploration, and yet little known outside Australia." Douglas Mawson, *The Home of the Blizzard* (1998, St. Martin's Press, New York).

Robert Rosenberg's 2005 article "Why Is Ice Slippery?" (December 2005, *Physics Today*, pp. 50–55) inspired a response by Vitaly Kresin describing work done in 1891 by the renowned experimental physicist Robert Wood. Wood put a block of ice in a powerful hydraulic press to demonstrate that pressure would not melt the water, arguing against what was then called the pressure-molten theory. Nevertheless, the belief that pressure from skis and ice skates melts the underlying ice remains alive today, even in textbooks.

In addition to being recognized as the world's northernmost tree, the dahurian larch is also long-lived. One dahurian larch in Yakutia, in Siberia, is believed to be more than nine hundred years old.

The diminutive willows, as they are sometimes known collectively, may be quite old despite their size. Individuals of 180 and 236 years old have been reported from Greenland.

Peter Marchand's *Life in the Cold* (1996, University Press of New England, Hanover, NH) provides very useful descriptions of plant adaptations to cold, including tables of temperature tolerances. In addition, Marchand provides information on human tolerance of cold in a chapter called "Humans in Cold Places." Similarly, James Halfpenny and Roy Ozanne's *Winter: An Ecological Handbook* (1989, Johnson Books, Boulder, CO) provides useful information in a chapter titled "People and Winter."

What may have been Alaska's northernmost accessible tree of any size grew next to the Dalton Highway, which runs parallel to the Trans Alaska Pipeline and connects Fairbanks to the North Slope oil fields. A prominent sign with large blue letters was erected in front of the tree saying FARTHEST NORTH SPRUCE TREE ON THE ALASKAN PIPELINE. DO NOT CUT. Perhaps inevitably, someone girdled the tree, apparently with the sort of small hatchet often carried by campers, ultimately killing it. There are, however, a number of smaller black spruce trees farther north just off the highway. They are not protected by signs and may eventually outgrow their protected but now dead neighbor.

Bernie Karl, the current owner of Chena Hot Springs Resort, is an outspoken entrepreneur. In addition to running his Aurora Ice Museum, he experiments with geothermal power, including generation of power using relatively low-temperature (less than two hundred degrees) hot spring water. He also heats his greenhouse with water from a hot spring.

Charles Darwin's description of events in Tierra del Fuego comes from *The Voyage of the* Beagle, first published by Henry Colburn in 1839 but reprinted many times and readily available as a Penguin Classic (1989, London). Readers of Darwin's *On the Origin of Species* will find a different Darwin in *The Voyage of the* Beagle. Darwin the scientist and thinker is still very much present, but he is accompanied by Darwin the seasick adventurer, a much more interesting and charming narrator.

The quotation about frostbite is from an e-medicine online clinical reference called "Frostbite," written by C. Crawford Mechem of the University of Pennsylvania School of Medicine (www.emedicine.com/emerg/topic209.htm). However, bearing in mind the history of race relations, one cannot dismiss the possibility that some of the differences seen in some studies and statistics reflect biases in the treatment of the research subjects. For example, it may not be possible to know whether black soldiers in Korea were given similar training, issued similar gear and food, and sent on similar missions as white soldiers. The most grievous example of racism in the annals of cold research comes from Nazi experiments at Dachau. This "research" is ignored here for ethical reasons.

Most of the information on the University of Alaska experiment comes from Ned Rozell's article "The Skinny on Humans and Cold" (February 6, 1997, *The Alaska Science Forum,* no. 1323). Rozell works for the University of Alaska's Geophysical Institute, translating arcane research into interesting articles that appear in numerous publications. Laurence Irving's book *Arctic Life of Birds and Mammals, Including Man* (1972, Springer-Verlag, New York), which includes a description of his experiment, is out of print but can be found in some collections.

Charles Wright's interviewer was Charles Neider, who edited the book *Antarctica: Firsthand Accounts of Exploration and Endurance* (2000, Cooper Square Press, New York).

DECEMBER

True Chinook winds are a type of foehn wind—a wind whose temperature increases as it moves down the downwind side of a mountain range. The temperature increase is caused by an increase in pressure with loss of altitude. Santa Ana winds are another type of foehn wind.

Kenneth Risenhoover collected data from January through April at Denali National Park as part of a study published in 1986, "Winter Activity Patterns of Moose in Interior Alaska" (*Journal of Wildlife Management*, vol. 50, pp. 727–34). In winter, the park is beautiful but brutally cold. One cannot help but wonder whether Risenhoover compared his experiences to those of Apsley Cherry-Garrard and his companions during their quest for penguin eggs.

The description of the use of a hot potato to assess the insulative qualities of a flying squirrel's nest comes from Bernd Heinrich's *Winter World* (2003, HarperCollins, New York). Part of the joy of reading Heinrich's books comes from his innovative approach to ecological research. Whereas some scientists would contrive complex field or laboratory measurements or experiments to measure the insulative qualities of nests, Heinrich used a hot potato, a watch, and a thermometer. Another wonderful aspect of Heinrich's work is his use of quotations from difficult-to-find publications. For example, both quotations about crossbill nest insulation later in this chapter are from *Winter World*. The 1900 quotation is attributed to J. Grinnell's "Birds of the Kotzebue Sound Region" (*Pacific Coast Avifauna*, no. 1), and the 1909 quotation is attributed to J. Macoun's *Catalogue of Canadian Birds*.

Many researchers have reported on the diversity of invertebrates found in snow, or the subnivean invertebrates. C. W. Aitchison, for example, has published a number of papers on invertebrate diversity in snow, including a series of papers in 1978 and 1979 in the *Symposia of the Zoological Society of London*, the *Manitoba Entomologist*, and *Pedobiologia*. A partial compilation of Aitchison's findings provided the list of invertebrates found in snow in Canada.

William Pruitt's book *Wild Harmony: The Cycle of Life in the Northern Forest* (1988, Douglas and McIntyre, Vancouver) translates the science of the Taiga into elegant and readable prose. Pruitt is also known for work in Alaska that contributed to the end of the U.S. government's

Project Chariot, which called for the use of hydrogen bombs to dig a harbor near Point Hope. Pruitt and his colleagues working on Project Chariot undertook what is sometimes described as the first environmental assessment ever to stop a major project. Dan O'Neill's Alaskan classic *The Firecracker Boys* (1994, St. Martin's Press, New York) describes Pruitt's role and its consequences.

Bethany Leigh Grenald's article "Women Divers of Japan" (Summer 1998, *Michigan Today*) gives one of the few contemporary accounts of ama divers in English. In an age of feminism and environmentally sustainable practices, it is difficult to understand why ama divers do not attract more attention.

The Bowhead Whale, a special publication of the Society for Marine Mammology, edited by John Burns, Jerome Montague, and Cleveland Cowles (1993, Allen Press, Lawrence, KS), is an excellent resource on bowhead whales. *Encyclopedia of Marine Mammals,* edited by William Perrin, Bernd Wursig, and J. G. M. Thewissen (2002, Academic Press, New York), is an extensive general resource on whales, including cold-water adaptations and feeding habits. The scientific literature on whales is remarkably vast and complex, even though these animals spend most of their time underwater in remote parts of the world's oceans.

The writing of J. Michael Yates is so lyrical that it blurs the distinction between prose and poetry. The passages here come from Yates's story "The Hunter Who Loses His Human Scent," first published in *Man in the Glass Octopus* (1970, Sono Nis Press, Vancouver). During a varied career, Yates worked as a prison guard in Canada, an experience he captured in his book *Line Screw* (1993, McClelland and Stewart, Toronto).

Apsley Cherry-Garrard's *The Worst Journey in the World* (2000, Carroll and Graf, New York) tells the story of the recovery of the penguin eggs. Despite the book's very appropriate title, Cherry-Garrard seems to have relished the journey, at least after the fact. One can imagine him enjoying the cold even as it beat him into submission. His two companions, Edward "Bill" Wilson and Birdie Bowers, survived the penguin egg journey only to die with Scott on the South Pole expedition.

Chicken, Alaska, with a current population of about twenty residents, originated as a gold mining settlement around 1880. The town's early inhabitants wanted to name their community Ptarmigan, because these chicken-like white birds were abundant in the surrounding country, but they could not agree on the admittedly odd spelling of this species. With the pragmatism of miners everywhere, they named the town Chicken instead.

David Sibley's *The Sibley Guide to Bird Life and Behavior* (2001, Alfred A. Knopf, New York) summarizes the scientific literature on feathers, including the evolution of feathers.

JANUARY

The hydrologists were working on the North Slope Lakes Project, described at http://www.uaf.edu/water/projects/nsl/reports/L9312_Hydro _Note_091906.pdf.

Arctic foxes are remarkably adept at finding food in and around human camps and facilities. Begging, foraging in supposedly animal-proof Dumpsters, and even brazenly walking through open kitchen doors are all common practices. Once animals are habituated to humans, the potential for aggressive behavior increases. Rabies, of course, makes aggressive behavior even more likely.

When a volume of air shrinks because it is compressed, the energy carried in the original volume of air is concentrated in the smaller volume, so the temperature increases. This is called adiabatic or isocaloric heating. The opposite effect is seen when pressure is reduced. Diesel engines, for example, use adiabatic heating instead of spark plugs to ignite fuel. A piston compresses air and diesel fuel until the temperature reaches the point of ignition.

The Coriolis effect is sometimes incorrectly called the Coriolis force. The spinning of the earth creates the appearance of a force, but no force is applied.

Most of the folk sayings about weather in this chapter are from a delightful small book by René Chaboud titled *Weather: Drama of the Heavens* (1996, Harry N. Abrams, New York). In addition to concise explanations about weather patterns and phenomena, the book presents dozens of color photographs depicting weather events, historical figures, and people collecting weather data from locations ranging from Siberia to Texas.

David Laskin's *The Children's Blizzard* (2004, Harper Perennial, New York) provides far more details about the Blizzard of 1888.

Lewis Fry Richardson's *Weather Prediction by Numerical Process* (1922; reprint, 2007, Cambridge University Press, Cambridge) was not his only accomplishment. Richardson also tried to use numerical analyses to understand the causes of war and is considered the cofounder of the scientific analysis of conflict, described in part in his *Arms and Insecurity* and *Statistics of Deadly Quarrels,* neither of which is currently readily available. Separately, he showed that shoreline length is a function of scale and that shoreline length will increase as finer scales are used. This so-called Richardson effect is often ignored in technical papers attempting to relate shoreline length to various ecological phenomena and political or economic statistics.

Some sources suggest that Lorenz had planned to mention a seagull's wings rather than a butterfly's wings. Lorenz was swayed to the butterfly by another meteorologist, Philip Merilees. The ideas expressed in the talk—intended to explain why accurate weather forecasting is so challenging—led to a blossoming of the much-misunderstood chaos theory, popularized in books and movies, including *Jurassic Park*. The essence of chaos theory is that small differences in initial conditions can result in huge differences in subsequent outcomes. Lorenz died on April 16, 2008, at age ninety.

James Glaisher wrote a full account of his balloon ascent, published on September 5, 1862, as "Greatest Height Ever Reached" in the *British Association Report* (1862, pp. 383–85). During the ascent, Glaisher describes himself fading in and out of consciousness at high altitudes.

He relates that his assistant, Mr. Coxwell, "felt insensibility coming over himself; that he became anxious to open the valve, but in consequence of his having lost the use of his hands [because of the cold] he could not, and ultimately did so by seizing the cord with his teeth, and dipping his head two or three times, until the balloon took a decided turn downwards." The account is available at www.1902encyclopedia .com/A/AER/aeronautics-33.html.

The anecdotes about strange weather events were found in *Weird Weather,* a collection compiled by Paul Simons (1997, Warner Books, London).

Part of Barrow's history is captured in Charles Brower's *Fifty Years Below Zero* (1942, Dodd, Mead, New York). The book describes in detail the mingling of cultures—the New England whalers and the Inupiat, who were well established thousands of years before the New Englanders arrived and began slaughtering bowhead whales. The Brower family name lives on in Barrow, with many families tracing lineage to Brower himself and an entire portion of Barrow known as Browerville.

The abundance of wildlife in the North Slope oil fields is likely caused by a combination of habitat availability, food availability (that is, from Dumpsters, workers ignoring restrictions on feeding wildlife, and other sources), and restrictions on hunting and trapping.

Arctic backcountry travelers often argue over the need to carry firearms for bear protection. In one often told story, an Arctic biologist dissuaded a curious polar bear by hitting it in the head with his shotgun. Many bear biologists consider the polar bear to be less threatening than its cousin the grizzly bear. Although weapons provide psychological comfort, it is worth noting that in addition to the risk of firearm accidents, shotguns add considerable weight to a traveler's supplies. In some cases, that weight capacity might be better used for extra food, fuel, shelter, or clothing or for a satellite telephone.

The bearded seal, *ugruk* in Iñupiaq, is a large ice seal—that is, a seal species dependent on sea ice for its survival. The spring whale hunt of the Inupiat, conducted from the sea ice, relies on the use of *umiaq*, skin boats made from *ugruk* skins, because the boats are light enough to drag across the ice in search of openings where whales can be found. By the autumn hunt, the nearshore sea ice has melted, and more substantial boats made from aluminum or fiberglass are used to search for whales in the open sea.

The 1881 expedition to Barrow was undertaken concurrently with Greely's disastrous trip as part of the first International Polar Year. A second International Polar Year was held in 1932–33, and a third International Polar Year was held from 2007 to March 2009. The International Polar Years are intended to focus scientific effort (and funding) on the Arctic and Antarctic.

Editor David Norton's *Fifty More Years Below Zero* (2001, Arctic Institute of North America, Fairbanks) provides a history of Western science in Barrow through the turn of the twenty-first century. That history continues to evolve rapidly today. The Barrow Arctic Science Consortium provides extensive logistical support for visiting scientists, and the North Slope Borough's Department of Wildlife Management supports and encourages both basic and applied studies.

Work by Nathan Pamperin, Erich Follmann, and Brian Person used satellite collars to track arctic foxes as they moved across the sea ice. One fox stayed on the ice for 156 days and traveled more than fifteen hundred miles without touching shore.

FEBRUARY

Tom Shachtman's *Absolute Zero and the Conquest of Cold* (1999, Mariner Books, New York) provides an interesting and plainly written account of the history of cold temperature research. It also introduced me to Frederic Tudor, who is surprisingly poorly known even in the Boston area. A biography titled "Frederic Tudor Ice King," published in 1933 in the *Proceedings of the Massachusetts Historical Society*, is available at http://www.iceharvestingusa.com/Frederic%20Tudor% 20Ice%20King.html.

Thomas Moore's *An Essay on the Most Eligible Construction of Ice-Houses; also, A Description of the Newly Invented Machine Called the Refrigerator* (1803, Bonsal and Niles, Baltimore) is available at http://www.digitalpresence.com/histarch/library/moor1803.html.

Giambattista della Porta's *Natural Magick* was first published in 1558 in Latin, but it was translated into Italian, French, and Dutch within a few years. It is available in English at http://homepages.tscnet.com/omardi/jportat5.html.

Bacon's speculations about Drebbel's use of saltpeter appeared in *Novum Organum*, published in 1620. A 2000 version of Bacon's work, edited and translated by Peter Urback and John Gibson, is available from Open Court Publishing (Chicago).

By the graces of a kind of magic never imagined by the likes of Drebbel and King James I, *Daemonologie*, originally published in 1597, is available in full at http://watch.pair.com/daemon.html.

Carl Wieman's quotation comparing a hailstorm to the use of lasers to slow molecular motion comes from a 2001 National Institute of Standards and Technology news release titled "Bose-Einstein Condensate: A New Form of Matter," available at http://www.nist.gov/public_affairs/releases/BEC_background.htm.

Erin Biba describes Lene Vestergaard Hau's work in which a beam of light was stopped cold in her article "Harvard Physicist Plays Magician with the Speed of Light" (October 23, 2007, *Wired*, www.wired.com/science/discoveries/magazine/15-11/st_alphageek).

MARCH

Among the many reasonably accessible discussions of frostbite is one by James O'Connell, Denise Petrella, and Richard Regan called "Accidental Hypothermia and Frostbite: Cold-Related Conditions," in *The Health Care of Homeless Persons—Part II—Accidental Hypothermia and Frostbite,* http://www.nhchc.org/Hypothermia.pdf.

The word "angora" is also said to have come from the Turkish city Ankara. Its roots can be traced to the Greek *ankylos,* for "bend," but perhaps because of its association with various animals, it came to mean "soft" in other languages.

Numerous Web sites provide summary descriptions about the processing of wool. Two examples are http://www.oldandsold.com/articles04/textiles13.shtml and http://library.thinkquest.org/C004179/wool.htm.

A more complete narrative describing World War II training intended to prepare troops for winter conditions in Japan can be found in "The Wet-Cold Clothing Team," published in the *Quartermaster Review* (January–February 1946), available at http://www.qmmuseum.lee .army.mil/WWII/wet_cold.htm.

There is no concise answer to the question "What is the best fabric for outdoor use in cold environments?" Different fabrics, including the many different synthetic fabrics, have different characteristics. One fabric may be warmer than another in the absence of wind but useless when the wind blows, another may be very warm until it traps moisture, and a third may be warm but unable to withstand day-to-day use. Differences are further obscured by manufacturers' claims, the absence of meaningful standard tests of warmth and durability, and the propensity for retail clerks to present themselves as experts. Making an informed choice about the best fabric for a parka or other winter clothing is as difficult as filing a federal tax return. Having said that, Hal Weiss's *Secrets of Warmth* (1992, CloudCap, Seattle) provides useful but dated guidance.

Thousands of patent descriptions can be found online by searching for patent numbers.

Although Vilhjalmur Stefansson's classic work *The Friendly Arctic* (1921, Macmillan, New York) has not been reprinted recently, copies of various old editions are available. The entire book also is available at http://books.google.com/books?id=zTvyrKu8PjwC&printsec=toc&dq =the+friendly+arctic&source=gbs_summary_r&cad=0#PPP1,M1.

Native American Niomi Panikpakuttuk's description of clothing comes from a Northwest Territories Archives transcription of a 1996 interview (document G93-009, Northwest Territories Department of Culture and Communications, Cultural Affairs Division, Oral Traditions Contribution Program, Yellowknife, NT).

Major and minor misconceptions about the Arctic are common and extend well beyond igloos. For example, well-educated people still believe that lemmings routinely form massive herds and march over cliffs.

Diamond Jenness's *The Indians of Canada* (1932, University of Toronto Press, Toronto), which has been reprinted several times, is still considered an important resource for understanding the history and culture of the native peoples of Canada. It includes chapters on hunting, dwellings, trade, social organization, religion, and other aspects of life. Jenness was born in New Zealand but spent more than thirty years trying to understand the native peoples of Canada before retiring in 1947.

For anyone passing through Fairbanks, the Cold Climate Housing Research Center is worth a visit. For information, see http://www.cchrc.org/.

APRIL

For more quotations that were misattributed to Twain, see http://www.snopes.com/quotes/twain.asp.

Robert Ettinger's *The Prospect of Immortality* (1964, Doubleday, New York) was published in English, French, German, Dutch, Russian, and Italian. Who would not be intrigued by the prospect of living forever, or at least longer than a natural life span? From the cover of the Doubleday English version: "Most of us breathing now have a good chance of physical life after death—a sober, scientific probability of revival and rejuvenation of our frozen bodies."

The brownish pages of a mimeographed copy of the cryonics manual have been scanned and made available at http://www.lifepact.com/mm/mrm000.htm by Fred and Linda Chamberlain, life members of the Cryonics Institute. In their introductory page, the authors describe the manual: "Notwithstanding this failure to 'get off the ground,' in the attempt to be comprehensive, a great number of topics were addressed, at least in a preliminary way." Although it may be easy to make light of cryonics, one can secretly hope that the Chamberlains and other believers will have the last laugh.

Roald Amundsen's *The South Pole: An Account of the Norwegian Expedition in the* Fram, *1910–1912* was republished by White Star Publishers (Vercelli, Italy) in 2007 as *Race to the South Pole*. Electronic versions of the original are available at books.google.com.

The quotation about potholes from the Washington State official came from "Recent Storms Leave Lasting Effects for Seattle Drivers," an article by Tiffany Wan in *The Daily of the University of Washington* (January 24, 2007). The quotation about potholes from a Michigan spokesman came from "Road Workers Scramble to Fix Winter's Damage," an article by Andy Henion in the *Detroit News* (March 13, 2007).

According to the Smart Road Web site (http://www.virginiadot.org/projects/constsal-smartrdoverview.asp), a 5.7-mile stretch of the heavily engineered and instrumented Smart Road between Interstate 81 and Blacksburg, Virginia, will eventually open to the public.

Earl Brown's *Alcan Trailblazers: Alaska Highway's Forgotten Heroes* (2005, Autumn Images, Fort Nelson, BC) is one of several books available on the history of the Alcan. *Alcan Trailblazers* relies in part on diary entries and letters written by construction workers. An interesting history with photographs can be found at http://web.mst.edu/~rogersda/umrcourses/ge342/Alcan%20Highway-revised.pdf. An *American Experience* documentary, "Building the Alaska Highway" (PBS), also presents the history of the road.

At least one document held by the Alaska State Libraries Historical Collections suggests that there was some interest in attempting to maintain the Hickel Highway (Alaska Department of Highway Photograph Collection, Walter Hickel ["Haul Road"] Construction, 1968–1969, PCA 82). Today many Alaskans, especially those working on the North Slope, refer to the Dalton Highway (the existing paved road between Fairbanks and Deadhorse that leads to the North Slope oil fields) as the Haul Road, not knowing that the Hickel Highway was also called the Haul Road. The routes for the two roads were not the same. Surprisingly few Alaskans have heard of the short-lived Hickel Highway.

Anchorage, although it is a relatively small city surrounded by undeveloped land and the sea, has had difficulties meeting federal clean air standards. For a full description of the situation, see "Clean Air Act Reclassification; Anchorage, Alaska, Carbon Monoxide Nonattainment Area" (December 2, 1997, *Federal Register,* vol. 62, no. 231).

Richard Byrd's experience with chronic carbon monoxide poisoning during his solo adventure in Antarctica is described in the August chapter. In his circumstances, levels could easily have risen to a point at which death was inevitable.

My house was flooded by a broken pipe while I was away overnight. When I returned, a distressing stream of water was flowing under the garage door, and the water in the downstairs rooms was ankle-deep.

Independence Mine State Historical Park is open to the public during the summer. Many of the mining buildings have been restored. Although visitors can walk around the grounds and hike into the surrounding mountains, none of the tunnels is open to visitors.

MAY

Although it seems unlikely that anyone can say for certain, most specialists seem to think that the carrying capacity of the Arctic steppe would have been somewhat less than that of temperate and tropical large-mammal havens today, such as the Serengeti.

Dan O'Neill's *The Last Giant of Beringia: The Mystery of the Bering Land Bridge* (2004, Basic Books, New York) tells the story of the scientific investigation of Beringia, focusing mainly on work by Dave Hopkins.

Postglacial rebound, also called continental rebound, isostatic rebound, isostatic adjustment, and isostatic recovery, is complex. As glaciers and ice sheets melt, land rises quickly in what is sometimes called elastic rebound. Later, the rate of rise decreases exponentially. Rebound rates today may be one inch every two or three years—difficult to measure but nevertheless rapid by geological standards. Rebound rates are measured using various methods, including surveying methods that rely on sophisticated GPS networks.

Evelyn C. Pielou's *After the Ice Age: The Return of Life to Glaciated North America* (1991, University of Chicago Press, Chicago) gives a much more detailed and technical account of many of the events associated with the end of the last period of extensive glaciation.

Kettle lakes are common in Alaska and throughout the north, but they are often mixed with other lake types, such as thaw lakes and oxbow lakes.

Plant and animal ranges continue to change, with the plants' and animals' entry into new areas assisted and accelerated by human corridors. For example, roads provide disturbed ground that allows certain plants to set seed with limited competition from long-established species that occupy most of the ground in undisturbed areas. Similarly, roads provide paths for animal movement. Radio-collared animals, such as wolves and caribou, often follow roads.

The first site where Clovis artifacts were found and the nearby Blackwater Draw Museum can be visited just outside Portales, New Mexico. When the Blackwater Draw Site was found during highway construction in 1932, bones were displayed in a Portales store as a curiosity. Edgar Howard, an archaeologist who was intrigued when he was presented with a fluted spear point by a resident of Clovis, New Mexico, excavated the site from 1932 to 1936 and referred to it as "the Clovis Site." (It was renamed the Blackwater Draw Site by E. H. Sellards many years later.) No human remains were discovered, but a number of stone and bone tools were found, along with bones from mammoths, shovel-toothed mastodons, ancient bison, horses (species that pre-dated the Spanish introduction of European horses to North America), tapirs, camels, llamas, dire wolves, ground sloths, short-faced bears, and saber-toothed tigers.

The scientists interviewed for the newspaper article about their paper on mosquito evolution were Christina Holzapfel and William Bradshaw, both of the University of Oregon. The two scientists published the arti-cle "Evolutionary Response to Rapid Climate Change" in the prestigious academic journal *Science* (June 9, 2006, vol. 312, pp. 1477–78). Among other things, the *Science* article said, "Studies show that over the recent decades, climate change has led to heritable genetic changes in pop-ulations as diverse as birds, squirrels, and mosquitoes." However, the authors also point out that an ability to evolve in response to rapid cli-mate change "does not, in itself, ensure that a population will survive."

Richard Stone's *Mammoth: The Resurrection of an Ice Age Giant* (2001, Perseus Publishing, Cambridge, MA) offers a thorough and readable history of the recovery of mammoth bones and frozen carcasses. Stone to some degree highlights investigations into the possible rebirth of the mammoth through the use of cloning technology applied to tissue sam-ples recovered from permafrost.

The quotation from Eugene Pfizenmayer, one of the scientists sent by the Russian Imperial Academy of Science to retrieve the frozen mammoth carcass from the bank of the Berezovka River in 1901–1902, comes from Pfizenmayer's *Siberian Man and Mammoth* (1939, Blackie and Son, London). This book describes both the reality of digging up mammoth remains and the rigors of traveling and living in Siberia in the early part of the twentieth century. Pfizenmayer's trip provides a contrast to Bering's trip across Siberia two hundred years earlier.

The Discovery Channel produced two documentaries about the Jarkov mammoth, *Raising the Mammoth* (2000) and *Land of the Mammoth* (2000).

A well-known North Slope wildlife biologist also commented on the freezer-burned taste of preserved meat during a public radio interview after tasting centuries-old frozen whale meat discovered in a long-abandoned ice cellar.

Northstar Island, also known as Seal Island, is a man-made gravel island in water about thirty feet deep six miles north of Prudhoe Bay.

Elisha Kent Kane described the expedition in *Arctic Explorations: The Second Grinnell Expedition in Search of Sir John Franklin, 1853, 54, 55* (reprint, 1996, Lakeside Press, Chicago). Kane provides yet another account of the fortitude required to survive multiyear strandings in the Arctic. Without intending to do so, he also looks somewhat foolish in his inability to learn from the local people who lived near his stranded vessel. Greely carried Kane's writings, among others, on his disastrous expedition in 1881.

The 1984 expedition to find the remains of Sir John Franklin and his crew is described in Owen Beattie and John Geiger's *Frozen in Time: Unlocking the Secrets of the Doomed 1845 Arctic Expedition* (1990, Plume Printing, New York). The book includes color photographs of the disinterred bodies of Petty Officer John Torrington and Able Seaman John Hartnell, both dead and frozen for more than a century but looking as if they could have been buried less than a week.

Brenda Fowler's *Iceman: Uncovering the Life and Times of a Prehistoric Man Found in an Alpine Glacier* (2001, University of Chicago Press, Chicago) provides a detailed account of the Iceman's discovery and exhumation, including the political machinations of various parties interested in this unusual find.

On October 15, 2004, Helmut Simon, one of the hikers who had discovered the Iceman, set out from Bad Hofgastein in Salzburg, Austria, on what should have been a four-hour hike. He did not return as planned. More than eighteen inches of snow had fallen. Simon died in the snow and cold of the mountains, just as the Iceman had thousands of years earlier.

Airman Leo Mustonen's story was told in dozens of newspapers, from Hawaii to Florida, suggesting the human-interest appeal of frozen human remains.

The museum housing the Iceman is in Bolzano, Italy. The Iceman's frozen carcass can be observed through a small window.

The dryas — specifically, *Dryas octopetala*, also called mountain avens — gave its name to two cold periods, or stadials, that occurred after the last glacial period of the Pleistocene Ice Age. The Younger Dryas, sometimes called the Big Freeze, lasted about thirteen hundred years, starting about thirteen thousand years ago. The Older Dryas lasted only a few hundred years, starting about fourteen thousand years ago. Stadials such as these (and the Little Ice Age) remind us of the variability of climate, but this variability should not be confused with the kind of variability that is occurring now, which appears to be much more significant and linked to greenhouse gas emissions.

JUNE

According to the Environmental Protection Agency, burning a gallon of gasoline releases almost twenty pounds of carbon dioxide. Many carbon footprint calculators are available on the Internet. The carbon footprint of even environmentally conscious people is shocking. No one would willingly and knowingly dump ten or twenty pounds (or more) of garbage from their car during the daily commute to and from work, yet that is exactly what most people do every day.

Joseph Fourier's 1827 essay "Mémoire sur les températures du globe terrestre et des espaces planétaires" ("Report on the Temperature of the Earth and Planetary Spaces," *Mémoires de l'Académie Royale des Sciences,* vol. 7, pp. 569–604) is often cited as the first description of the greenhouse effect. To some degree, it appears to be a rehashing or refinement of Fourier's 1824 paper "Remarques générales sur les températures du globe terrestre et des espaces planétaires" ("Remarks on the Temperature of the Earth and Planetary Spaces," *Annales de Chemie et de Physique*, vol. 27, pp. 136–67). His papers have to be read in the context of the times. One would not expect someone working in 1827 to have even a rudimentary understanding of the current knowledge of heat exchange and atmospheric physics, so readers should not expect Fourier's papers to offer any more than a hint of the truth as we understand it today.

Guy Callendar's first paper on climate change and carbon dioxide seems to have been "The Artificial Production of Carbon Dioxide and Its Influence on Temperature" (1938, *Quarterly Journal of the Royal Meteorological Society,* vol. 64, pp. 223–37).

Carbon dioxide dissolved in water is an acid. The impact of ocean acidification is only beginning to be understood, but it may turn out that acidification equals or exceeds climate change in its ability to disrupt ecosystems and affect the lives of humans.

A paper by Roger Revelle and Hans Suess titled "Carbon Dioxide Exchange Between Atmosphere and Ocean and the Question of an Increase of Atmospheric CO_2 During the Past Decades" (1957, *Tellus*, vol. 9, pp. 18–27) is often considered to be an especially significant paper in the development of current thinking about climate change.

On May 15, 2008, the polar bear was listed as "threatened" under the Endangered Species Act. The listing was driven primarily by loss of sea ice habitat. Under the act, a species is considered "endangered" if it is at risk of extinction in all or a substantial portion of its natural range in the foreseeable future and "threatened" if it is at risk of becoming "endangered" in the foreseeable future.

Bob Carter was quoted in the *Canada Free Press* in a June 12, 2006, article called "Scientists Respond to Gore's Warnings of Climate Catastrophe." The article goes on to say that most of Gore's climate change supporters are not climate change experts and that many climate change experts are not strong supporters of predictions of widespread and rapid climate change. Predictably, Carter was promptly attacked from some quarters and praised from others.

The quotation from Richard Lindzen comes from an August 1, 2006, article by Lindzen. The article was published by the Heartland Institute in *Environment and Climate News* under the title "No Climate Change." Lindzen describes Gore's vision as "shrill alarmism." Lindzen is widely cited for challenging claims of a "consensus" among scientists regarding climate change.

The Patrick Michaels quotation comes from a June 21, 2005, article by Ker Than called "How Global Warming Is Changing the Animal Kingdom" in *Live Science*. Michaels does not argue against climate change but rather points out that not every change in animal and plant communities is linked to or caused by climate change. "It's not all a result of human induced climate change," he is quoted as saying. "Half of it is at best, probably less than half."

The information about changes in how the frozen body of James Bedford was stored comes from an article called "Dear Dr. Bedford (and Those Who Will Care for You after I Do)," published in the July 1991 issue of *Cryonics,* available at http://www.alcor.org/Library/html/BedfordLetter.htm. The author of the article, Michael Darwin, is difficult to reach but is widely recognized as an important figure in the field of cryonics. The quotations from John Baust and Arthur Rowe come from the Alcor Institute's Web site. It is refreshing to see an organization involved in a controversial endeavor that is willing to quote its critics.

The Arthur C. Clarke quotation is from a letter written in support of a legal case on June 20, 1989. Although Clarke is perhaps best known for his science fiction, which includes *2001: A Space Odyssey* and *Childhood's End,* he seems to have seen endless possibilities coming from scientific advances. In his letter supporting the possibility of successfully preserving human life through cryonics, when he wrote, "Although no one can quantify the probability of cryonics working, I estimate it is at least 90%—and certainly <u>nobody</u> can say it is zero," Clarke underlined "nobody" and signed the letter "best wishes." The letter is available at http://www.alcor.org/Library/html/declarations.html.

The quotation from Bert Lenten is from an article in the electronic newsletter Mongabay.com. The quotation from Robert Hepworth is from the May 13, 2007, edition of the *International Herald Tribune (Europe)*.

The June 2006 issue of the National Park Service's magazine *Alaska Park Service* includes a series of photographs showing various Alaskan glaciers at different times, ranging back to the early twentieth century and continuing to more recent times.

Bill Ford Jr.'s comments about climate change were released with Ford Motor Company's 2005 climate change report.

In the late 1990s, the oil industry began to acknowledge publicly the possibility that climate change related to greenhouse gas emissions could be an important environmental problem. Under the leadership of Lord John Browne, British Petroleum (later officially renamed BP) was the first of the major oil companies to acknowledge climate change as a potentially important issue. Browne's announcement came in May 1997 during a speech at Stanford University, where he said, "There is now an effective consensus among the world's leading scientists and serious and well informed people outside the scientific community that there is a discernible human influence on the climate, and a link between the concentration of carbon dioxide and the increase in temperature." Browne's entire climate change speech is available at www.bp.com/genericarticle.do?categoryId=98&contentId=2000427.

The Arctic sea ice is changing rapidly. Satellite images or recent summaries based on satellite images should be checked for current information.

The comments from Sergei Kirpotin and David Viner come from "Thawing Siberian Peat Bog Will Speed Up Global Warming" in the August 11, 2005, *Moscow News*.

Sea ice near shore is often dirty, sometimes because it has been frozen to the sea bottom or the shore and has picked up sediment, but more often because river floodwaters run over the top of the ice or wind blows sediment or sediment-laden spray from the open water onto the ice.

INDEX

INDEX

Birdseye, Clarence, 173, 230
bison, 211–12, 217
Bjerknes, Vilhelm, 143
Blair, Tony, 240–41
blizzard of January 1888, 22–26, 38, 39, 63,
    143, 149, 230
blubber, 126–27, 128, 133
Bose, Satyendra Nath, 168, 169, 230
Bose-Einstein condensate, 168–69, 172,
    174, 230
Bowers, Birdie, 132–33, 267n
bowhead whales, 126–28, 154, 213, 237
Boyle, Robert, 162, 169
Brainard, David, 20, 243, 253n
building foundations, 6, 37, 38, 76, 170, 188,
    189
Burroughs, John, 81–82, 261n
Byrd, Richard E., 29, 43–44, 190, 254n,
    256n, 275n
Byron, Lord, 52–53

California, 115, 148, 206
Callendar, Guy, 231, 280n
calories: and hibernation, 46, 79; intake of,
    86–87, 117, 127; polar explorers' burning
    of, 20, 44–46; and rete mirabile, 129;
    and shivering, 20, 44, 81, 134, 135, 205;
    storing of, 205
Canada, 60, 120, 170, 179, 186–87, 233
Candlemas Day, 147, 156, 157
cannibalism, 19, 230
carbon dioxide: and airplanes, 228; and
    automobiles, 227, 228, 279n; buildup in
    houses, 189; and dry ice, 6, 165; freezing
    of, 146, 229; and global warming, 99, 229,
    230, 231, 232, 282n; levels of, 229, 230,
    231, 232, 282n; measurement of, 231;
    methane compared to, 241; in oceans,
    231, 280n; and Pleistocene Ice Age, 213,
    230; and refrigeration, 160; and subnivean
    dwellers, 121, 189
carbon footprint, 279n
carbon monoxide poisoning, 43–44, 195, 199,
    256n, 275n
caribou: and avalanches, 207; clothing from,
    183, 187; foraging of, 11, 121, 237;
    migration of, 11, 99, 238; size of, 98; as
    winter actives, 117, 118, 121, 122
Carothers, Wallace, 181
Carré, Ferdinand, 173
Carter, Bob, 233, 280–81n
caterpillars: difficulties in finding, 9, 10, 11, 19,
    47; as food for explorers, 19; freezing in
    winter, 6, 11; pet caterpillars, 49, 66–67,
    74, 83, 124, 194, 195, 217, 234
caves, 40–41
Celsius, Anders, 7, 8, 165
Celsius scale, 6, 7, 8, 250n

Challenger (space shuttle), 198–99
Chamberlain, Fred, 274n
Chamberlain, Linda, 274n
Champlain Sea, 213–14
Channeled Scablands, 69
chaos theory, 144, 268n
Chardonnet, Hilaire de, Count, 180
Charpentier, Jean de, 62
Chena Hot Springs Resort, 107–09
Cherry-Garrard, Apsley: on degrees of frost,
    8–9, 70, 251n; on frostbite treatment, 39;
    on ice hell, 9, 12, 15; and penguins, 132–
    33, 266–67n; on relative temperatures,
    124; and Scott's expedition, 12
China, 32, 89, 159, 218
chinook winds, 115, 119, 121, 158, 264–65n
Clarke, Arthur C., 235, 281–82n
climate change: causes of, 37, 228, 229, 230,
    231, 232, 233–34, 238, 259n, 280–81n,
    282n; climate variability and predictability,
    257n; effects of, 236, 237–38; and genetic
    changes, 277n; reconstruction of past
    climates, 67–72; and sunspot activity, 55.
    See also global warming; Little Ice Age
clothing: and adaptation to cold, 110–11, 112,
    113; and chemical heat packs, 176, 182;
    of Clovis people, 215–16; and cotton, 176,
    178, 180, 203; and dampness, 57, 73, 123,
    175, 181, 203, 228; and fur, 183, 187, 216,
    224; and gold mining, 203; of Iceman of
    the Alps, 224; and insulation, 175, 178,
    181; layering of, 178–79, 183, 204; and
    midsummer cold, 22; and paradoxical
    undressing, 26; and rescue of German
    troops, 4; and synthetics, 176, 180–82,
    184, 203, 272n; in temperate zone, 85;
    and underwater diving, 112, 122–23; wear
    and care of, 29, 178; and weight capacity,
    269n; and windchill, 25, 58, 181; and
    wool, 177–79, 180, 203; and World War
    II, 151
Clovis people, 215–17, 276–77n
Cockburn, Alexander, 238
Cold Climate Housing Research Center,
    189–90
Collins, Jake, 73–74
Comcock (Inuit), 14–15
Cook, Frederick, 15, 44, 99, 253n, 257n
Cook Inlet, 156–57, 158
corals, 68, 96, 115–16
Coriolis, Gaspard Gustave de, 141
Coriolis effect, 141, 151, 267n
Cornell, Eric, 169, 172, 251n
cotton clothing, 176, 178, 180, 203
cranes, 87, 236
Crimean War, 142–43
Croll, James, 63–64, 65
crops, 205–06, 232, 241

286

ABOUT THE AUTHOR

BILL STREEVER manages an environmental studies program in Alaska and serves on many related committees, including a climate change advisory panel. He lives with his partner and son in Anchorage, where he hikes, bikes, camps, scuba dives, and cross-country skis.